A DECENT
BOTTLE OF WINE
IN
CHINA

Chris Ruffle

EARNSHAW
BOOKS

A Decent Bottle of Wine in China
By Chris Ruffle

ISBN-13: 978-988-82737-2-0

TRAVEL/ Asia/ China

First printing November 2015

EB073

Published by Earnshaw Books Ltd. (Hong Kong)

This is the story of how we came to set up Treaty Port Vineyards, and build a Scottish Castle accompanied by fifty acres of grape trellises on the north coast of the Shandong peninsula in eastern China. I started writing with the idea of producing a sunny travelogue, along the lines of *A Year in Provence*, but along the way things took a darker turn. I have drawn a veil over some of the difficulties and tried to accentuate the affirmative but have not always succeeded, as you will see. I hope this story provides a taste of what we had to do to produce our first bottle of decent wine, whilst, on the way, showing a little of the changes underway in Chinese society.

Chris Ruffle, October 2015
www.treatyport.com

The superior man knows that the world is imperfect,
but does not lose heart.
Confucius

Contents

CHAPTER I

IT WAS MAY DAY

The desire for safety stands against every
great and noble enterprise.
Tacitus

IT WAS MAY Day in 2004. I had been visiting companies in
Taiyuan, the capital of gritty Shanxi province, a region rich in
coal, noodles and history. I was making these visits as part of
my day job, investing in China's stock markets for a British
fund management company. My wife, Chang Ti-fang (hereafter
known as Tiffany), had accompanied me to take advantage of
the long weekend to visit some tourist sites. Our prime target
was Pingyao, a well-preserved old town still surrounded by its
Ming dynasty walls and well worth visiting. I had heard from
a colleague that in the vicinity there was a vineyard, which
employed a French winemaker and produced drinkable wine.

I was, at this point, a wine novice. I enjoyed drinking wine,
and had visited some vineyards when I worked in Australia,
but knew nothing beyond this. My few encounters with grape
wine in China, usually a desperate attempt to avoid the more
lethal Chinese vodka-like spirit, *"baijiu"*, had not been pleasant.

I had once organised a barbecue for a group of investors on the slopes of Laoshan near Qingdao at the Huadong Winery, which had produced a passable Chardonnay. More typical, however, was an American colleague, and wine snob, who, when asked by me whether he would like to try a Chinese wine to go with his Chinese meal answered: "Ask me in a hundred years..."

So it was in a spirit of adventure that we left the main road, and bumped our way along the track to Grace Winery. The place itself - large white-painted concrete buildings - lacked charm, and the dusty Shanxi countryside was brown and uninteresting. But the wine was good, especially the Merlot. Over lunch with the winemaker, Gerard Colin, we tasted (and I was charged for) four bottles. Gerard is large, stockily-built and as bald as a potato. He smoked steadily throughout our meeting and said, were he to stop smoking, it would take him a while to re-calibrate his taste buds. He explained how, now over 60 years old, he had come from his home in St. Emilion to this remote and lonely spot (a long and rather sad story which involved him losing his father's property in a failed business venture and finding himself with a large financial hole to fill). We talked about rugby union, popular in his part of France. I have a picture of our meeting that first day, with Gerard looking strangely uncomfortable.

Surely Gerard should be able to find a more pleasant and accessible site for a vineyard in all his travels around China? He agreed that there was just such as place, a lovely valley by a lake near the coastal resort of Penglai in Shandong. He mentioned that it should be possible to set up a small vineyard there for just US$1 million. He had a French acquaintance, a Monsieur Humbert, who worked for the Yantai government and might be able to help. The seed was planted (and Gerard's first wild underestimate made).

One month later, I was visiting companies in Shandong, an Eastern seaboard province with a population of about 100 million. I stayed in Yantai, a large port on the north coast of the Shandong peninsula, which sticks out towards Korea, as evidenced by the many Korean companies which have now set up there. Yantai is also the home of the Changyu Pioneer Wine, founded in the early 20th century by Zhang Bishi, an overseas Chinese businessman from Java, with help on winemaking from the Austro-Hungarian diplomat Baron Max von Babo. His initial contract promised $200 per month plus lodging and a share of the profits, but forbad him from telling anyone if the wine was no good. The company is today listed on the Chinese stock market, was a few years ago the subject of a management buy-out and is now market leader. Its museum and restaurant are well worth a visit, but I suggest you drink tea with your meal (if you must order Changyu wine, you should order the white from their recently-purchased New Zealand property).

I made time to visit Yantai Hill, now a pleasant park dotted with consulates built by the major foreign powers after Yantai was opened as a treaty port in the 1860's (it was then known to foreigners as Chefoo). Each nation's consulate is built in its own national style, with Britain's colonnaded consulate building occupying, of course, the best position, overlooking the harbour. The old American consulate contains a small, but worthwhile, museum describing Yantai's history over this period. It is all described in Marxist terms – foreign colonialists exploiting Chinese workers in the semi-feudal society of the time – but it looked to me more like foreigners creating local employment through trade and investment, often for little reward.

As I was later to learn, Yantai's existence owes much to the first English consul, a Mr. Morrison, who decided that its port was larger and deeper than the originally nominated treaty port of Penglai (old name, Dengzhou). After all these years, all the wars and revolutions, I was impressed to find that there is still a Morrison Street in Yantai.

I decided to visit the "lovely valley" that Gerard had recommended and contacted Mr. Humbert, a rather eccentric Frenchman, retired from an expatriate career in chemicals, who preferred to stay on in China to help the local government attract foreign investors rather than return to his native shores. Together with his colleague, Jack Xia, we drove for about one hour in a minibus, ending with a steep climb up a muddy track through apple and peach orchards. I do not remember much about my first short visit. But the weather was good when we arrived-- somehow weather in Qiushan valley often seems better than in the surrounding area. The scenery was, indeed, lovely. Coming over the rise, the valley suddenly opens out in front of you, with the rocky slopes of Qiushan on your left, and a long view down to a shining lake in the distance. The hillside reminded me of a recent holiday in the castle-dotted mountains of Northern Spain, its granite bones jutting out from beneath the fresh green summer grasses, scattered with low pines and cypress. I have spent so much of my time in China in anonymous, grey cities, visiting ugly factories, it was a delight to find such a place.

My next trip to Shandong was that August. Tiffany saw the site for the first time. Gerard also came, accompanied by a long-haired photographer-turned-architect called Gao Bo. We met the local officials of nearby Daxindian town, together with representatives from Penglai's foreign trade department tasked with encouraging inward investment. We lined two sides of a large table in the dingy town hall and drank tea with the leaves floating on the top (the key tactic when drinking is to blow before you suck). The mayor of Daxindian deferred to the party secretary and his deputy, an early indication that politics in China does not work as in the West.

We were shown a detailed proposal for the valley to be turned into a golf course. This had fortunately been scuppered by the central government's recent ban on further golf course developments (I'm with the Communists and Mark Twain on this one). It was on this map that we first saw marked the memorial to the Taoist scholar Qiu Chuji, which attracts thousands of worshippers, and the site of his temple, which had been demolished after the communist victory. Qiushan reservoir, created in the 1950s by damming the valley, is the major source of drinking water for Penglai city, so no industrial developments are possible in the vicinity. We outlined our idea for developing high quality vineyards in the area, and emphasised the need for sensitive development in the area and protection of the environment. There was earnest nodding.

We ate the first of many banquets with a selection of officials. Heavy eating and drinking are a key element of doing business in China. Inhibitions are broken down as the table fills with plates and alcoholic toasts multiply. Soon everyone is an "old friend". Fortunately the seafood and vegetables of Shandong are delicious and cheap, although some of the more exotic dishes can try the palate of the most adventurous Westerner. Donkey, dog and insect larvae are all local favourites; the first two are passable but I advise giving a wide berth to the larvae.

This is the first time we climbed to the top of Qiushan, smelt the wild thyme sprouting from the crumbling granite soil and, whilst recovering our breath, saw the hawks launch themselves from the summit and hover in the breeze. On the precipitous north side of the hill, which is unsuitable for cultivation, there is woodland, which I am sure acts as a nature reserve to the local area. We climbed across an old wall near the top, which turned out to be the remnants of fortifications built by (or to defend against?) local Taiping rebels, the "Nien Army", in the mid 19th century. We munched fresh-picked apples and apricots, generously offered by the local farmers. Gerard crunched his way through a turnip, whittling away its skin with his penknife. We looked for possible sites for a winery and eventually chose the spot where the contours and view best suited.

It was Gao Bo who came up with the plan to build a road from the lake, rather than to go with the government's rather insensitive ("poor fengshui") plan to carve a road across the facing hill. The government, to our pleasure, agreed. Little did we foresee that this would not be a scenic winding country road, gradually revealing glimpses of the castle, but a straight highway; the authorities do not know how to build roads in any other fashion. And that the original road would, anyway, be built as planned. China has more than its share of such "pork barrel" projects. It was also on this visit that Gerard spotted a courtyard of tumble-down farm buildings, once used to grind peanuts, on the edge of the local village, Mulangou ("Peony Ditch"). He felt that these premises might be useful as an office and store whilst the winery was being built.

On this visit we met two future employees for the first time. Gerard introduced Dr. Guo Donglin, who had returned after several years of study in France to work for the Penglai Trade Bureau. Dr. Guo, with many contacts in the wine business, became our consultant and subsequently helped us to navigate our way through the labyrinth of local bureaucracy. We also met "Old Huang", the village head of Mulangou, heavily-built, dressed in army fatigues, as we wandered around the ruined farm buildings. This might have been the first time we heard his catch phrase (always said with a broad smile on his weather-beaten face at moments of adversity) "*Mei wenti*", "No problem". This did not actually mean that the problem would be overcome, but it made everyone feel better.

Back in Shanghai there followed an exchange of e-mails and faxes with the Daxindian government regarding terms of a contract. A local lawyer was employed. Party Secretary Liu and Mme Zhang Li from the Foreign Trade Bureau came to our 40th floor apartment in Shanghai for dinner. This accelerated negotiations, again proving the lubricating affect of wine and good food on business. At one point Liu produced a list of the taxes we were expected to pay.

"But I don't think we need to bother with this one," he said, and leaned over to scribble it out with a ball-point pen.

This was repeated as more wine was drunk until about one half of the list had been scratched off. The search also started for an architect to design the castle/winery. Gao Bo showed no interest (too far from Beijing, no carte blanche on design) and an American architectural firm said they could build me ten wineries, but could not build just one.

I recall sitting awake in the stale air of a long flight over the Pacific, whilst others slept around me. I took out the notebook I use for recording company visit notes, and started to sketch what a winery on the slopes of Qiushan might look like. I drew a castle, which would make use of the gradient of the slope. I did have some experience in this area, as you will see from the next chapter, so this is not quite as strange as it might seem. And, all boys like castles.

This is my e-mail to the Scottish architect Ian Begg, dated the 29th September 2004:

A DECENT BOTTLE OF WINE IN CHINA

Dear Ian,

I hope that you are well. I am not sure whether you will remember me - you helped me to persuade Historic Scotland that Dairsie Castle could be re-built (much against their better judgement). I think it turned out pretty well in the end.

I now live and work in Shanghai. I have a project to grow the best wine in China, in joint venture with a friend who is a wine maker from St. Emilion. The vineyard is near the seaside town of Penglai on China's Shandong peninsula (the bit that sticks out towards Korea/Japan - please consult your atlas). As the "chateau" for the vineyard, I would like to build a Scottish castle. It is on a South-facing slope, so my idea is to have the wine factory and cellar under a courtyard, onto which the castle would face. The grapes will be hand-picked and sorted, and the wine made using gravity, with no resort to pumps, so the verticality of the castle would have a practical use. As well as hosting the tasting and selling of our wine, the castle will also be used for the wedding trade. The project also includes the restoration of a courtyard house in the local village, will be used to house the manager and visiting guests.

I remember you once said that you would like to build a Scottish castle in California. Well, this is not California, but it is a beautiful spot, looking down towards a lake over a valley full of orchards of apples, peaches and apricots. On the top of the hill where the castle would stand (called Qiushan) there is actually the remains of an old fortification dating from the Taiping Rebellion. It is our long-term plan to encourage first-grade wine makers from several countries to come to set up vineyards around the lake. There would be no Historic Scotland to contend with - the local government officials, eager to encourage our investment, are marvellously flexible on planning.

My idea is that you might consider coming up with a design after visiting the site. You could then work with a local design institute on its implementation. Local labour and materials are cheap (there is actually a granite mine not far away). My budget for the architectural design is £40,000 plus expenses.

I realise that this all sounds rather eccentric, but let me know if you have any interest and I will tell you more.

Regards,
Chris Ruffle

Ah, so idealistic, so naïve….

CHAPTER 2

REBUILDING DAIRSIE CASTLE

MY INVOLVEMENT WITH China was not the result of any far-sighted plan, but merely of youthful rebellion. Tired of being asked what I wanted to study at university, I picked the strangest subjects I could think of: Chinese with Philosophy. I am from Bradford in Yorkshire with no relative who had ever been to Asia, apart from grandfather, who won a cup for swimming across Hong Kong harbor when he was a private in the army. My father, an engineer frustrated in selling to European companies by his lack of foreign language skills, had always encouraged me to learn languages, but I am not sure Chinese was quite what he had in mind. And so I ended up studying Chinese at Oxford, one of only four students in my year in the whole university, in the late 1970s. The subject was taught with enthusiasm (to any of my teachers who happen to read this, my thanks). It did not seem at the time to be a terribly practical course, but it is possible that knowledge of Neo-Confucianism in the 12th Century and of the guerilla struggle in China between 1937 and 1945 has served me quite well in my subsequent career.

I must thank the Bradford Chamber of Commerce for being the first to get me to China. This upstanding body offers a scholarship to help local students improve their knowledge of foreign languages, so as to boost British trade. As a young man, faced with an imposing panel of Bradfordian businessmen, I was told, 'At this point in the interview we normally bring in someone to test your oral standard. But as we were unable to find anyone who speaks Chinese, perhaps you'd just like to say something.' After I'd burbled on for a few sentences, the chairman said, 'Well, that seems all right to me', and gave me the scholarship. Gentlemen, I hope you consider your money well spent.

My link with China has survived from 1977, when I opened my first Chinese-language textbook, written in the Gang of Four era, to the present day. On graduating in 1981 I could find no employer interested in anyone speaking Chinese, so I ended up selling Fairy Toilet Soap in Newcastle-upon-Tyne. However, by 1983 I got to Beijing with a metal trading firm called Wogen Resources, and then opened their Shanghai office in 1984. I have spent most of my time since then in Asia including stints in Taipei (1990–93 and 2000–02), Hong Kong (1993–94) and Shanghai (2002 to the present), with many visits in between. I got into finance by learning Japanese and taking an excursion into the Japanese stock market bubble in the late '80's. This financed the restoration of a Scottish castle in the 1990's, of which more below. My travels have taken me to all the provinces in China, with the sole exception of Hainan.

I now live on the unfashionable side of the river in Shanghai with my wife and youngest daughter (when I first lived in Shanghai, the only reason to take the ferry over to Pudong was to be able to take a photograph of the Bund). Here, I run a fund management company called Open Door (www.odfund.com) which invests in China's entrepreneurial companies, mostly for the benefit of institutions in Europe and the U.S. Except, that is, on many weekends in spring and autumn, when I am to be found in Shandong…

In 1992, I bought the ruin of Dairsie Castle in Scotland. I was then working in Taiwan for a company called S. G. Warburg. At that time, I did not own any property, and friends told me I ought to get on the property ladder in London. I looked at some houses in Chiswick, an area in West London which I got to know when studying for a thesis at the Public Records Office in Kew. But everything looked terribly expensive and, at that time, I happened to read a small article in the paper about a man restoring a castle in Scotland. This sounded a lot more interesting and started me on the quest for a suitable ruin to rebuild. Of course, financially, I would have done far better with the house in Chiswick – but then you would be reading another story.

I remember that first project, the one I had read about in the paper, was called Glasclune. It turned out to be a nice spot (and the only place I have ever seen a red squirrel) but impossibly remote and completely impractical for anyone with a day job.

This is the case with most unrestored castle ruins in Scotland. So I registered myself with various estate agents in Scotland as a man interested in buying a castle ruin - "Frivolous property" as a lady at Savills called it; I imagined ruins being filed in her cabinet under "F" , alongside those Scottish islands that come up for sale every summer. I had some clients in the financial community in Edinburgh and decided I should be able to persuade them to employ me at the appropriate time. So, being a practical type (why are you smiling?), I used a compass to draw a circle representing a one hour commute from Edinburgh. I then restricted myself to looking at candidates within the circle.

As I was working abroad, it was my poor parents who bore most of the burden of checking out the prospects. Some ruins were next to an industrial estate or motorway; one was in the middle of a messy farmyard. For another castle, the brochure photographer had managed to exclude enormous mounds of tar oil tailings which surrounded it. One prospect in the town of Penicuik south of Edinburgh dominated the town in a way which, in this day and age, is socially unacceptable ; my parents found a petition in the local butchers protesting potential redevelopment of the site, which was, judging by evidence, much used by courting couples and glue sniffers. We also found that you don't want your ruined castle to be too old; medieval castles tended to be temporary refuges with tall, thick walls and tiny windows, unsuitable for a modern home. The most restorable castles come from the mid 16th century, when the local lairds got their hands on church money, with the dissolution of the monasteries, and decided to build themselves a chateau – more a fashion statement than the grim fortifications of yore.

Dairsie Castle came up as one of the lots in a farm sale. I knew we had a real candidate when my mother described it as "not bad" (in Yorkshire, this counts as lavish praise). This ruin came with several advantages. It is a only 15 minutes drive from the famous university, and golfing town of St. Andrews, and only 10 minutes from Cupar, which has a station on the main line to Edinburgh (the eventual door-to-door commute turned out to be 90 minutes). The ruin stood in six acres on a bluff above the River Eden, at the head of the fine Dura Den valley, next to a pretty little church. On the downside, there was not much left of the ruin. Also, on the aforementioned railway, about 500 metres away, trains wooshed by every hour or so. I won the lot with a bid of £50,024 (now you know my lucky number).

As a Grade II listed ruin, Dairsie came under the supervision of Historic Scotland. Our first meeting with this august body did not go well; they "did not view Dairsie as a suitable candidate for restoration." As I discovered, official policy towards Scotland's ruins had undergone a sea change. In the 1970's and '80's the restoration of historic ruins was not only encouraged, but a number of people I met had received public funding towards such restorations; such people included Sir David Steel, former head of the Liberal Party, Sir Nigel Fairburn, Conservative MP and Member of the European Parliament... are you seeing any pattern yet? Some such publicly-funded restorations had been undertaken in concrete breeze block or brick, disguised under harl (a lime-based coating). By 1992, however, it was all about authenticity and maintaining the historic fabric "as was". Now in theory this is fine, but unfortunately ruins, untended, tend to fall down. It is also difficult sometimes to tell what "authentic" means, as any house which has been lived in over several hundred years tends to have had changes made to it, so is the 18th century addition less "authentic" than the 17th century addition?

To compete, I realized I would need to learn a lot more about Scottish history and architecture. I used subsequent holidays visiting Scottish Castles and those who had restored them, as well as lobbying members of the Buildings committee. This was generally an enjoyable task – people who have gone through the process of castle restoration tend to be keen to share their experience – though it also showed up some of the challenges. Should I get the chance to rebuild Dairsie, I determined that it would be to make it a comfortable, modern family home, with proper heating, plumbing, insulation, car parking and an elevator (more of the latter anon).

Our breakthrough came from an unexpected direction. I sponsored an archaeological dig around Dairsie, led by a lady from the archaeology department at St. Andrews University called Edwina Proudfoot. This stalwart lady, tall with short grey hair and always, in my mind's eye at least, carrying an umbrella, proved more than a match for Historic Scotland. Not much was discovered in the excavation of interest to non-archaeologists – some Yorkshire pottery, some Venetian glass, some clay pipes, a carved water spout – but it helped to delineate the foundations of the castle. It showed that the original east-west block had an added entrance tower at the north-east corner, as well as the two round towers at the north-west and south-west corners. I had chosen Ian Begg as my architect. He was a veteran of a number of Scottish castle restorations, so I thought would be able to hold his own with Historic Scotland. Also he had built, and was living in, his own entirely new castle at Plockton on the West Coast, so I thought he would be aware of the requirements of modern life. It was he that discovered in the National Gallery of Scotland a water colour of the castle by the antiquarian Captain Francis Grose from about 1780, which shows the eastern end of the castle still occupied, and the south-western tower covered by a conical roof, perhaps after its conversion to a doocot (where they kept doves, for eggs and meat). This provided us with sufficient information to make an educated guess as to what Dairsie would have looked like. Within one year of purchase we had permission to rebuild.

Unlike my later experience in Shandong, at Dairsie I used all direct labour to reconstruct the castle. An eager young man called Tim Heale, fresh from rebuilding a castle in the highlands, was my local representative; he helped to hire the individual workmen. One of the things that most impressed me about the Dairsie experience was how we were able to find locally so many skilled craftsmen; chief amongst them were the stonemasons Dod McArthur and Stan Walker. The former, who lived in a caravan onsite for much of the project, suffered from epilepsy, which made his walks across the high scaffolding worrying for informed onlookers. Dod and Stan were helped by Alf Hill, with his dragon-tattooed arms and, for the fine carving of the various plaques and dormers, Martin Reilly. Seeing the amount of work needed, especially for the dressing of curved ashlar stones for the round towers, increased my admiration for the original masons who lacked our power tools. The local stone is a reddish-tan sandstone, which is easier to work than the granite used for the later castle in China.

Local mines were long closed, but I managed to buy a supply from the demolition of derelict Lathockar House. (Ironically an earlier owner of Dairsie called Sir George Morrison fought a duel against the owner of Lathockar House in 1661 – belated revenge?) Unfortunately we ran out of stone before we reached the square cap houses, which go on top of the two round towers, which is why they are a deeper red.

We also had an excellent blacksmith, David Wilson, for all the cast iron work, and good carpenters – Ronnie Leadbitter for the shuttering and Keith Potts for the doors and shutters. Keith's most famous exploit came when we discovered the large four-poster bed that I had had made would not fit through the windows of the main bedroom, so Keith took out a section of floor , finagled the bed through, and put it back without any remnants of the join. My only failure was the plumber, so he won't get a mention here. In China I did not find such craftsmen. Perhaps this is because I did not look hard enough. However, I suspect it has more to do with 90 percent of the building stock you see in China having been built in the last 30 years; there are not sufficient old buildings to allow such a cadre of craftsmen to survive. The overwhelming bulk of Chinese construction is about volume and speed, with little attention given to the quality of finish. Cement and brick are the materials *du jour*, and the Chinese builders were visibly uncomfortable with stone. The roof of Dairsie was originally probably stone slab, but we were able to persuade Historic Scotland to allow us to use Caithness slate, which the local slaters, Niven & Sons, laid with a minimum of fuss. Slates are laid by nailing them through a drilled hole to an underlying wood structure in an overlapping fishscale formation. We explained this so the Chinese builders, who we discovered, after they had been paid and gone, had merely cemented on the slates. Which is why the Scottish roof remains sound, but slates are sliding off the Chinese version already.

Perhaps I will revert to the clay tiles with which the Chinese builders are more familiar.

The structure of the Dairsie castle rebuild is quite clever. Where it is new build, and not the massively thick stone walls of the original, it was decided to use an internal skeleton of cement onto which the outer stone skin was held by stainless steel ties. This left an internal gap, which provides excellent insulation, and a place to hide all the down pipes, whilst making the modern walls feel sufficiently thick. The only place we did not use this in Dairsie, at the top of the tiny curved stair by the northeastern caphouse – the result of geometries not working out – is the coldest in the house. The same design was employed in China, where it should have been easier because there was no ancient fabric to integrate. But the architects kept trying to bridge it with cement and, in the end, the builders just filled in the gap with rubble.

Which is why there are rain water pipes running down the outside of the castle in China, but not at Dairsie.

Historic Scotland kept a beady eye on us throughout the

rebuild. I can remember a controversy about whether we would be allowed a back-door, which was only achieved through the intervention of the fire service. As for the elevator, we hatched a plan to install it in the space left by the fireplace and flue from the old kitchen. I can remember the inspector, a bespectacled chap called Richard Fawcett with a small furry moustache, refusing on the basis that "most people don't have lifts in their homes". I forget how we got this one through (vital given the number of winding spiral stairs, which were the alternative). I was taken to court on one matter, "the use of re-constituted stone in certain window lintels". All existing windows were repaired in stone but, in the absence of one penny of public assistance, and with finances tight, I decided to use reconstituted stone for all new window lintels. Apparently I was reported by Sir Nigel, the previous recipient of government largesse, who saw the blocks lying in the field, but I defy anyone to tell the difference now. I was fined £250 in Cupar sheriff's court, slightly more than the man in the previous case who had head-butted his mother, but saved £60,000, which is what it would have cost me to carve all the lintels in stone.

For the last year of the rebuilding process I found a job with a fund management company in Edinburgh and moved back to Scotland. We lived in Pitcorthie House, an impressive modern plate glass and cement house built on the site of a much earlier pile, and occupied by the now Lord Lindsay and his Swiss wife, before he acceded to the title. We were once invited by our landlords to Balcarres, the family seat, for tea.

We arrived, all togged out in our Sunday best, and were given a nice tour of the property. The Flemish carvings in the dining room are particularly impressive. Before we knew it, we were shown out of the back door sans tea, leaving my parents muttering darkly about Scottish hospitality. We had had a similar experience when Sir Nigel invited us to lunch at his castle. We arrived to find the wife gardening; as I advanced across the lawn, bearing flowers, I could see the realization flash across her face, "Shit, we invited them for lunch". So we did not get lunch either. We were invited back, when we were treated to a couple of hours of Sir Nigel's excoriating wit. When he died a couple of years later I saw a local TV interview with his wife; she was asked what she thought of the discovery of Sir Nigel's love-child in Australia, and coolly answered "It helped to accelerate the grieving process."

Being back in Scotland gave me time to complete my researches about Dairsie, a process which I enjoyed, putting flesh on the bare bones. The castle has some interesting history. A Scottish parliament was held at Dairsie in 1335, probably because the regents for young David II felt it was a safe distance from the invading army of of Edward III. Sir James Learmonth, treasurer to James V and early convert to the protestant cause, had acquired the castle in 1517 and died at the slaughter of the Battle of Pinkie in 1547. His son Sir Patrick held his place as Provost of St. Andrews for 45 years. He played an important role in the Lords of Congregation army which faced down the Queen's French troops on nearby Cupar Muir in 1559. In 1575 he rescued Lord John Hamilton, leader of the Queen's party, from an ambush (allegiances seem to have been flexible in those days). This event, which prompted the only siege in the castle's history, is commemorated in a painting which hangs in the Great Hall, painted by a descendant of Michael Lermontov. This famous poet is himself a descendant of the Dairsie Learmonths, who ended up fighting in Russia as mercenaries, and being ennobled. The young James VI also found shelter at Dairsie when escaping from captivity at Ruthven.

John Spottiswoode (1565–1637) acquired the castle in 1616 one year after he had been made Archbishop of St. Andrews. This well-travelled and learned man was the main representative for the Stuart kings in Scotland, given the thankless task of re-introducing bishops to the Presbyterian church. Several of his writings survive, including his magnum opus "The History of the Church of Scotland" for which King James told him to "Tell the truth, man, and spare not". When he died he was given an 800 torch burial in Westminster Abbey, rather than, as specified in his will, the quiet burial alongside his wife in the church he built near the castle.

Both his son Robert, Secretary of State for Charles I, and grandson John, were brought to the scaffold through their association with the mercurial Royalist general Montrose, a frequent visitor to Dairsie[1].

During my research, I became convinced that the interiors of 17th century castles would have been very different from the gloomy ruins we normally visit today. I believe they were brightly painted to impress and, where money allowed, hung with tapestries or panelling for comfort. Furniture would have been more sparse than in the stuffed Victorian baronial castles and would be the light tan of freshly worked oak, rather than the time-blackened pieces now on display. So this is the look that I aimed for. I was lucky to find an array of talented artists to realize my dreams. Robert Koenig carved the main doors and shutters on the piano nobile level, as well as the oak tree sculpture in the garden. Martin Rayner carved the columns for the minstrel's gallery, and the newel posts at the top of the three spiral stairs sometimes, terrifyingly, with a chain saw. He had my

---

[1] There is a more detailed history at www.dairsiecastle.com/history

older children help him paint the flames of hell, welcoming the fall of Lucifer and an Intercity 125 train, at the base of the column. Jenny Merridew painted the entrance roof, the stair murals and the Hokusai bath.

Dairsie is brighter than the castle in China, where the linen-fold panelling I found has a darkening effect relative to Dairsie's white painted walls. Also, at Dairsie, I am indebted to George Begemann's lighting design.

Whilst we were working on the restoration, I learned that we could also restore the barony. In Scotland, the title goes with the ownership of the castle, rather than heredity. The Lord Lyon, the Queen's representative in matters of heraldry, asked me what symbol I would like for my coat of arms. I thought a Chinese dragon would be appropriate, given my long career in China but, lacking artwork, sent him a newspaper cutting of the dragon from a Dragon Airlines advertisement. It was Lord Lyon who suggested the dragon grasp an oak tree, the symbol of the Spottiswoodes.

In the Treaty Port Vineyards crest, I have substituted a glass of red wine. But the motto remains the same: "Unitas Scientiae et Actionis". This is a translation into Latin of the most famous precept of my favourite Chinese philosopher (and Neo-Confucian) Wang Yangming. In English, it is rendered as "Unity of Knowledge and Action". Wang, unlike many philosophers, seems to have been a man of action; his concept of the Unity of Knowledge and Action states that you should not do something until you know what you are doing, but once you know, you should do it.

The plaque above the front doors of both Dairsie and the castle in Shandong reads "Dilexi decorum domus Tuae" ("I have seen the beauty of Thy house"). This comes from Psalms 26.8, and was chosen by Spottiswoode for the plaque above the door of Dairsie church. I have a Renny Tait painting of how Dairsie now looks in the landscape, next to Spottiswoode's church and a 1950's concrete silo on the road to Cupar, built to hold sugar-beet.

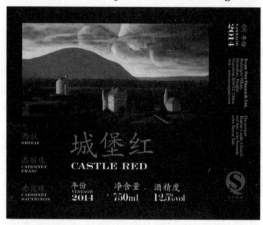

As we rebuilt Dairsie, the field in front became something of a wasteland of stones and thistles; I remember once finding a flock of goldfinches feasting on the thistle heads. After completion, there was a question of what to do with this waste. I had them bulldoze it into a curving berm, now grass covered, which looks like an ancient fortification or a ramp used to besiege the castle. When we moved in, the castle was still covered in scaffolding and the truculent removal company refused to carry our boxes in, offloading them into the field, leaving the family and builders to take them inside. I remember my grandfather's heavy pianola causing the most trouble. Apart from swimming across harbours, he was a self-taught pianist, playing the pianola with his eight workable fingers.

We received our completion certificate on March 15th 1996, one day before the castle warming party.

It is difficult to compare the two castles. I lived at Dairsie castle for five years, before moving back to China and I have included in the castle in Shandong a number of the features which I found most successful, such as underfloor heating, an elevator, a herb garden and murals. But Dairsie was a family home, whilst the castle in China is both a factory and hotel, often busy with visitors to be entertained or officials to be schmoozed. There is rarely time to relax in Shandong, but when I get a chance I enjoy the library, which I think is the most successful room in the castle. On a fine day, there is also a spot under the wisteria in the front garden where I am difficult to find. The weather at the Chinese castle is certainly much better, and there are a lot of plants that can be grown here but not in Scotland (grapes, of course, but also basil, sunflowers, tomatoes and peaches). At Dairsie, the breeze in spring is flavoured by wild garlic which thrives in the woods; on the hill above the castle in China, the scent is of wild thyme. At both places you can hear the rasping call of pheasants, which were originally imported from China to Britain as game for shooting. I remember once seeing a fine looking pheasant in a cage at the Dundee zoo that challenged Darwin's theory of evolution; on the plaque it stated that this was the type of pheasant originally imported into the U.K., but it was found to fly too fast and too high, so it was dropped in favour of a slower bird.

The stars can be bright at Dairsie, but you could only see them by braving the winter cold; in the Scottish summer, when it might be comfortable to star gaze, the nights are too light. The summers at Dairsie were nice, but commuting in winter was not much fun; I have a strong memory of the black, wind-swept platform at Cupar waiting for the 7am cross-country train. In China, however, I will always be the foreigner, "the boss", whereas in Scotland I was merely mildly eccentric (and so one amongst many).

CHAPTER 3
THE DEAL IS DONE

WE NOW JUMP back to Shandong, where in October 2004, Ian, now 81 and looking ever more like Gandalf, arrived with his wife, Ruth. This time Qiushan was climbed on a stormy grey day and the site surveyed. The design for the castle-cum-winery which Ian sketched on his first visit is pretty much how it was built.

Three rather bemused local construction companies had been lined up for discussions with Ian. Even with pictures and photos, communicating the concept was difficult. Perhaps this should not be a surprise in a country where construction is measured in cubic metres of reinforced concrete and high design is cladding this concrete in bathroom tiles paired with blue, reflective glass.

It was at this meeting that details of the contract with the government were ironed out. The basic idea is that we rent 200 *mu* (about 14 hectares) for the vineyard and buy the 10 *mu* on which the castle is to be built. In China, all land is owned by "the people", so what you are buying is only the right to use the land for a specific period of time, residential land for seventy years, industrial for fifty years and agricultural thirty years. No one knows what will happen at the end of the specified periods, but if there is not some ability to roll-over ownership, I guess there will be another revolution. Wine is, slightly oddly, categorized as industrial. In return for the investment, the government promised to provide road access, plus water, electricity and phone connections to the site. The capital of US$1.5 million was to be payable over four years.

These terms were not arrived at without much haggling; the party secretary and his deputy, playing bad cop good cop would come up with some outrageous term (e.g.. pay all 50 years rent up front), only then to deny they had meant any such thing when we raised a cry. Anyway, contracts are viewed differently in China; they are merely an expression of what both parties think, or hope, at a particular moment, and can be reinterpreted or ignored when circumstances change or prove inconvenient. As the deal was signed, eating and drinking reached a crescendo; at one disastrous meal toasts were made in a *baijiu* which tasted like paint thinner, a local grape wine, and a bottle of whisky which Ian had brought along. This was all topped off with a tonic wine with a snake curled inside. The consequences were horrible, if predictable. No one can say I haven't suffered for my art.

The copy of the map we received at this time delineating the vineyard bears almost no similarity to the vineyard as finally planted. The reality is that the compensation which the government offers to the farmers is just a fraction of what I paid to the government. Consequently, many farmers refused to hand over their land and my vineyard is, to this day, dotted with patches of peanuts and corn. Also, the compensation which is paid to the farmers is greater if it is already planted with fruit trees. On more than one occasion, having viewed a potential plot, I have come back the next day to find it planted with three inch-high apple tree seedlings.

We also agreed terms with Gerard. Then he, the CFO Tiffany

and myself, the chairman, headed off to a local photo shop to get our pictures taken as directors of the new company "Treaty Port Vineyards (Penglai) Ltd". I chose the name as a link back to those European pioneers of a century before. The Chinese version (Denglong hongjiu) is more poetic and of Tiffany's devising: Deng is derived from the old name for Penglai (Dengzhou) but also sounds like "lamp" or "rising"; long means dragon, referring to the company symbol, and hongjiu means "red wine". We agreed with Old Huang to hire a bulldozer and workers to start terracing the land to be planted next spring and left confident that everything was underway. When we phoned a couple of weeks later, from a holiday to Sparta, we discovered that the government had prevented work from starting, being upset that we had decided to use local farmers for land forming, rather than the rude and expensive contractor they had suggested.

Work was eventually started late, meaning that our bulldozing was done in an early snowfall. In early December, I paid a solo visit to Penglai at the government's request. This was to attend a press briefing to announce all the foreign investment projects which Penglai had attracted. Sitting in the audience I listened to Penglai's party secretary announcing the roll of honour until finally he reached "the vineyard investment from Mauritius" (Treaty Port's parent company, Jam Tomorrow, is based on the lovely and tax efficient island of Mauritius). Apparently the investment value of our project was now US$9m! Turning to the mayor sitting next to me, I commented that inflation in his town seemed to be rather fierce! He gave an embarrassed smile. This number, a complete and heroic fabrication, is the one that will appear in the roster of "inward investment attracted" on which his promotion will be based. It will also contribute to the national statistics on which economists place so much reliance.

I have a photograph of Treaty Port's first office. It is in a broken-down village hall with a slogan for family planning still visible on the walls (we are still in the "one child policy" era). Someone has messily repainted the door a bright blue, and drips of paint run over the doorstep and splatter onto the neighbouring grey brick wall. The door is closed by padlock. A telephone wire has been jury-rigged through the top of the door frame. A sheet of A4 has been printed with the words "TREATY PORT VINEYARDS LTD. PENGLAI OFFICE" and affixed to the inside of the door window pane. From such small beginnings...

CHAPTER 4

SETBACKS NOTWITHSTANDING...

Very few things happen at the right time,
and the rest do not happen at all.
Herodotus

OUR NEXT VISIT to Penglai was in January 2005, accompanied by my Canadian relatives, Geoff and Lynda, fresh from attending my wedding (Tiffany and I had got married in Hong Kong, followed by wedding dinners with family and friends in Taipei, Shanghai and the Lake District – terribly complicated). Terracing work had been suspended as the ground was frozen and could not be re-started until March. The government had not been able to provide the 200 *mu* contracted. Some of the land was unsuitable; some was planted with Forestry Commission pine trees, and some land the farmers simply did not wish to sell to the government. Faced with the imminent need to decide on the number of vines to import, my aunt made the eminently sensible suggestion that we only plant half of the land in 2005. In the end, as it turned out, the land was ready, ploughed and expensively fertilised. The idle half of the land was then planted with soybeans to improve the soil.

We opened our local bank account and handed the government our first cheque in payment for the first year's rent and the disused farm buildings in Mulangou. Tiffany had to re-write that cheque four times for 1) Using the wrong type of pen 2) Using of the wrong type of Chinese characters--they should be the simplified characters used in the PRC, rather than the traditional characters still used in Tiffany's native Taiwan 3) Our seal being chopped in the wrong place and 4) The cheque being creased. Cheque writing is quite an art in China. All this was done and re-done in the freezing front hall of the local Bank of China branch, whilst the bank employees sheltered warm behind plate glass screens; customers do not come first in China's state-owned banking system.

Penglai Construction ("Pengjian") had shown themselves the most responsive of the local construction companies so in March we signed with them to rebuild the Mulangou farm complex in the local style. This would include an office and meeting room, a warehouse, four bedrooms with en-suite bathrooms and a kitchen, dining hall and laundry room. I was surprised by Pengjian's flexibility when it came to the contract negotiations. It only became clear later that we had grossly overpaid and left out a couple of crucial clauses. Anyway the contract was signed on March 30th and by April 1st there were 30 workers camped onsite. Here they remained until the rebuild was completed in October in conditions of no little squalor. Tiffany sent occasional food parcels to relieve the staple diet of cabbage and large bread dumplings Cans of Coca-Cola proved particularly popular.

The problems with construction in China are not the same as in the West. My experience in Scotland is that builders tend to take on too many projects, then leave you in the lurch and take forever to complete. In China, buildings go up quickly. The rafter-raising ceremony, when rockets and firecrackers were exploded and sweets thrown by workers on the roof to the crowd below, took place just five weeks from the start of construction. The problem in China is that builders can make construction decisions so bizarre that you cannot even conceive that they needed to be guarded against. On one visit I discovered that the door between the office and the toilet was clear glass. When I exasperatedly asked the builders whether it wouldn't be embarrassing for someone using the toilet to be watched by all their colleagues, and was met with a stare of blank incomprehension. Electrical sockets were installed directly below water faucets. Nice looking doors guided rain water directly into the house. The sanitary ware installed was so petite it could have been used in a doll's house. The list was long and its correction tedious. The drilling of a well also took a heroic effort, only reaching water 58 metres down.

A more serious potential debacle was the import of vines from France. Gerard proved to have been overconfident in his ability to import vines, bought from Guillaume, an old friend in Bordeaux, despite several warnings from better-informed locals.

Paperwork proved insufficient and we faced the prospect of missing the planting season or seeing our vine shoots drying to dust in some customs warehouse. Dr. Guo rode to the rescue. A minibus was filled with presents and set off on a road trip, gaining approvals from Yantai, the provincial capital Jinan and eventually the Forestry Ministry in Beijing, where the last present was handed over. The planting of vines began on May 20th, very late in the season. Against the odds, the vines grew pretty well in their first year. Only three percent were lost. Mandy Liu, Gerard's assistant, played an important role in measuring the vineyard and supervising planting and watering. She eventually married Gerard's son, Pierre, who Gerard tried to persuade us to hire on several occasions. I understand the happy couple opened a bar in Beijing.

I have a picture of the Treaty Port team in the fields on a typical Qiushan early summer evening. It is still not hot enough for us to be stripped down to T-shirts; the air in the valley is rarely still, and there is probably a fresh breeze blowing from the East. It is blowing back the rim of Tiffany's hat, which should be guarding her against a sun tan. The team here includes Mr. Ruan, the head of Qiushandian village near Qiushan Lake, which initially provided 90 percent of the land for the vineyard (nearly everyone in Qiushandian is called Ruan or Zhang). Shortly after hiring Old Huang, he was replaced as head of Mulangou village, by a younger character, also called Huang (everyone in Mulangou is called Huang or Gong). This hooligan, feeling left out of the action despite our best efforts, continued to make difficulties for us, making ill-founded complaints and looking for hand-outs. Some years later, I found him, older, sadder and hopefully wiser, working in our vineyard.

One day, Old Huang was knocked off his motorbike by a truck. Failure to wear a helmet (helmets are rarely worn in rural China) made his convalescence a long one; his only daughter worked in the city, and tried to provide some long distance assistance, as did we. This event highlighted for me the problems in rural healthcare. Where once Mao's barefoot doctors brought some limited assistance to farmers, now they had no insurance and an injury could impoverish entire family. This is one of the main reasons why Chinese people's savings are so high. When later the government realized it needed to boost consumption, and so lower savings, I knew it would need to address this problem, and was motivated to set up a China Healthcare Fund. Today, rural health insurance coverage, though only RMB 320 per head, is over 90 percent.

The next project was to install posts and wires to control next year's vine growth. We decided to use granite, rather than concrete, as it is more attractive and would match the castle walls. This project was undertaken by two teams under Mr. Huang (cousin of Old Huang) from Mulangou and Mr. Zhang from Qiushandian. Roughly half of the posts (2,200) were installed before early snow halted activity. We added granite panels to inform future visitors to the vineyard what grape type they were looking at.

With the frozen ground we now started to plan for 2006 and an early second planting to avoid a recurrence of the first year's debacle. This also entailed the first pruning which was scheduled for March.

The completed farm buildings were ready to be furnished. On a final survey, I noticed that plaster stars remained on the roof gables from the original peanut oil factory. I instructed the workers to paint it red (as it was, doubtless, in the original). When Tiffany saw this, she told me that her father, who had been forced to flee his home ahead of the Communist army, would never sleep in our house as long as it bore a red star. Visitors today, if they look closely, will notice that the farm has only white-painted stars. With the farm deeds to be delivered shortly and Pengjian paid, we were free to push ahead with the plans for the castle and winery, prepared by Ian Begg. Ian and Ruth visited Penglai again in June to discuss with the two remaining construction companies in detail how to adapt Ian's plans to Chinese regulations. In China, all foreign architects' plans must be interpreted by Chinese architects in the light of local regulations. I was shocked when I saw the first Chinese version of Ian's design, which seemed to have shrunk like a failed soufflé. This turned out to be a function of the minimum required width and rake of the main spiral staircase, around which a tower house is designed. This was eventually resolved, after much red ink, by moving to a scale-and-plat design for the main staircase. Ian also recommended bringing in Ove Arup to help with engineering work within the factory. The next step was to be detailed negotiation of the construction companies' quotes.

This summary of 2005 would not be complete without mention of the sterling efforts of new recruit Mrs. Lu, who worked diligently to chivvy the contractors and farmers to complete their allotted tasks. Her less ebullient husband, Mr. Liu, worked for Bank of China Penglai branch, and provided to be of considerable assistance, helping us cope with the arcane idiosyncrasies of China's banking system. The capitalisation of Treaty Ports was increased to US$1 million in preparation for 2006.

I had to ask Gerard Colin to resign as a director and manager of Treaty Ports in December, in a disagreement over expenses, conflicts of interest and time-keeping (though often charming, he wasn't terribly reliable). We started talking to potential replacements. It was also about this time that one female member of staff started calling Tiffany in the evening, drunk, to tell her that she loved her. Perhaps this shouldn't be a book but a soap-opera.

At Chinese New Year 2006, Ms Lu organised a gift of musical instruments to the villages of Mulangou and Qiushandian. As the gifts were drums and cymbals the term "musical instruments" is perhaps something of a misnomer. These instruments were used to accompany the village dance troupe, later dubbed "the Mulangou Ballet." The youngest participant appeared well into her 40's. Standing to view the performance, I tried to look interested and stamped my feet to keep the blood circulating. After both villages had performed, the head of the Mulangou team sidled up to me.

"We have prepared another dance. Would you like to see it?" he asked.

I had by now lost all feeling in my toes, but, being a sensitive chap, and not wanting to hurt his feelings, replied: "Well, okay, if it's not too long."

After murmured assurances, the team leader hurried off to rally his troops. More clanging, banging and orchestrated flag-waving ensues. I was staring at the clouds, wondering why the farmers stacked baskets of corn cobs on their roofs, when the noise suddenly stopped. I clapped, but not too loudly, as I was wearing gloves and didn't want to encourage an encore.

Then, of course, the head of the Qiushandian team, looking slightly put out, strode up.

"We have a second dance as well," he declared. "I am sure you would like to see it?"

I have never felt more like the Queen.

CHAPTER 5

A Change of Direction

Only by constant change can a man be constant in
wisdom and happiness.
Confucius

Change? Aren't things bad enough already?
Lord Palmerston

As if by magic, we now jump forwards to summer 2007. The
first chapters were written during the enforced idleness of the
Chinese New Year holiday 2006, all Chinese businesses being
closed and Tiffany's pregnancy preventing us from leaving
Shanghai for more pleasant surroundings. Now, I had a one-
year-old daughter, Eliza, who had already made several trips to
Shandong. I believe that she was conceived in Mulangou, but my
wife disagrees. She particularly likes cherries, the first local fruit
to be available, followed in season by apricots, peaches, grapes
and apples.

There had been some major changes to the vineyard. We
hired Christophe Koch, introduced to us by Dr. Guo, to manage
the technical aspects of the vineyard.

Christophe, with his wife Florence and six children (he didn't look that old), is based in Perpignan in the Southwest corner of France. He travelled to Mulangou for one week each month, but eventually planned to move his whole family out to China.

Christophe had a wide range of experience in the wine industry, acting latterly as advisor and salesman to a group of organic wine growers in the Southwest of France, where we visited him one summer to visit vine suppliers and Cathar castles. He became a feature of the Qiushan countryside, tearing round the vineyard on his racing motorbike with his translator, Miss Chen, riding pillion.

His first piece of advice was to cancel the plan to import vines for the remaining 100 *mu* of land. This was partly due to the lack of suitable vines - some of 2005's vines showed signs of having been fridge-stored, a problem due to a huge surplus of vine shoots in France. Also Christophe felt that the rocky terraces of Qiushan needed reshaping and deep ploughing. The aim was to produce larger plots, accessible to tractors, to make maintenance of the vineyard more efficient. The gap between rows was set at a minimum of 1.8 metres. A system of drainage ditches was dug to channel rainwater, occasionally heavy in mid-summer, into a variety of reservoirs. Hundreds of tons of fertilizer were ploughed into the land; I signed one cheque for 600 tons of chicken shit. At times it looked like a bad day on the Somme. Soybeans were once more planted on unused land and ploughed in. One overenthusiastic farmer even planted the car park with soy beans. A team of elders, led by our night-watchman (another Huang) used stones extracted from the fields to build some fine dry stone walls.

During 2006 the size of the vineyard grew. In addition to the unplanted 100 *mu* from 2004/2005, a further 100 *mu* were acquired, extending the vineyard onto the eastern slope of Qiushan. The new land included a small amount from the two villages of Xinxing and Sungshan. The aim was to increase the potential production of the vineyard to a more economic scale and to monopolise all available land in Qiushan valley. The acquisition of the land was a tricky three-way negotiation between the company, the local government and the farmers using the land. Deals were finally struck. We used our heavy equipment to create paths to farmers' fields previously difficult to access; reservoirs were dug for mutual benefit. But you cannot please all of the people all of the time and we had to involve government officials in a number of disputes.

Christophe also had to undertake a quick and surreptitious re-burial ceremony having accidentally dug up a coffin. In rural China there are no cemeteries as such, but graves of varying degrees of sophistication dot the landscape; south-facing slopes are preferred. Paper money was duly burned.

Work on the planted part of the vineyard, was divided evenly between Zhang and Huang, together with their families. I have a photograph of them discussing tactics over lunch with Qiushandian village head Ruan. That's not water in their glasses, which is why it's best to do business with Shandong farmers in the morning, if you want to understand what they are saying. Also, I have found that most accidents in the vineyard take place in the afternoon.

In the already planted part of the vineyard we made some mistakes. Other vineyards in the area prune their vines in January. This probably stems from the practice for pruning apple trees at that time. In France, the rule is to prune in March, after the threat of frost has passed. This seems to be the more sensible approach, but as February closed, we came under increasing pressure from local "experts" to prune.

We conceded, and the result, unguided by Christophe, set back the vineyard's development. The other mistake was a lack of attention to local conditions. In the South of France, the rain mostly falls in the winter and the summer is dry. The opposite is the case in Shandong, where July and August can be very wet. We are keen to take an organic approach to the vineyard, not using the large amount of chemicals loved by Chinese farmers. But the rain caused mildew problems on the vine leaves and we were slow to spray. The relatively narrow gap left between vines rows in the first planting means that spraying could only be done using the farmer's own "walking tractors" or, in the case of sulphur, on foot. The lack of deep ploughing for the first planting showed in relatively weak growth in some vines planted in rockier areas (notably the lower Viognier plot). We cut off all the green grapes which appeared in this the 2nd year in order to stimulate root growth, much to the disgust of the local farmers, who thought this was a terrible waste.

Much work was done trying to persuade the local government to make good on promises in the contract, such as building a road to the castle site, as well as providing water, electricity and communications, and planting trees to hide the tombs on the hillside. (The tombs didn't bother me, but apparently, as we hoped to hold weddings at the castle, these should be hidden). There was a setback when Liu, the party secretary of Daxindian, with whom we had spent much time eating and drinking to develop relations, was promoted to Penglai. The new party secretary, a rotund man called Zhong, was barely willing to give us the time of day. We were helped by a senior local politician, Liu Shuqi, who has been a major force behind development of the wine industry in Penglai. He invited us all, including Zhong, to a lunch in Penglai.

As host, Liu sat at the head of the table, facing the door; Tiffany and myself were either side of him; Zhong sat directly opposite. We had barely exchanged the opening pleasantries when Liu, a tall man, leaned forward across the table, and looking directly at Zhong, said: "Treaty Port Vineyard's investment is an extremely important project for the development of the wine industry in Penglai. I am keen that it receives full support from local officials."

"Yes, certainly," said Zhong, leaning away from the table.

"And I will be extremely upset," Liu continued, "if I hear of anything less than full co-operation. Are we quite clear?"

Zhong leaned even further from the table and examined his finger nails. "Absolutely."

Even then, it took considerable efforts, including help with a trip by Zhong to visit wineries in Australia, and a round of letters from me recommending his son to U.S. universities, to win his grudging compliance. This requirement for the constant maintenance of political *guanxi* (relationships) is taxing and something I never understood before I started this project. My admiration for successful businessmen in China increased.

My idea of a winding country lane up from the lake was swiftly disabused when, in May, the government built a two-lane road to the factory door which was concreted by the autumn. Another road was built over the hill from the south for no apparent reason, offending all the rules of *fengshui* (this later turned out to be linked to negotiations with Lafite). A further lane was constructed between our plots closest to Mulangou; this leads to the memorial to the 13th century Taoist philosopher Qiu Chuji, confidante of Gengis Khan, after whom the hill and lake are named. The temple that had been knocked down in the first flush of revolutionary fervour was due to be reconstructed. After further breakthroughs on the "*guanxi*" front, the highway, rather than stopping at the castle, was in May 2007 connected to the village of Mulangou. In a sudden flush of environmental enthusiasm, the government also lined this road with alternating trees and conifers. It was not initially possible to tell from these white-painted, leafless tree stumps what type of trees they were, but those that survived the truck traffic turned out to produce pink blossom in springtime, which is not only pretty, but can be eaten in dumplings.

Unfortunately, the thick conifers have created a number of blind bends, to the detriment of speeding traffic and pedestrians.

Considerable time was spent in 2006 planning what vines to plant in 2007 and how to source them. The trick was to find the right varieties with the most appropriate rootstocks, and also to ensure that they were fresh. Due to overproduction in France, many vines are now stored in fridges for more than one season, weakening the plant.

An additional obstacle was that only a limited number of pepinieres are licensed to export plants to China. In the end, we decided to split demand for the 85,000 plants needed between two suppliers, the Bordeaux firm of Arrivé, which has a joint venture nursery in Penglai, and Provence-based Dayde. Vine quantities were determined by the amount of each variety needed to fill tank sizes chosen for the winery. We added to our existing quantities of Syrah, Cabernet Sauvignon and Merlot. Varieties new for 2007 were Grenache, which Christophe feels should thrive on the sunnier upper slopes, Marselan (a new Cabernet/ Grenache hybrid resistant to mildew) and Chardonnay, for acidity.

The planting in 2007 was finished during April, but not before several alarums which seem to be an inherent part of getting anything through Chinese customs. My attempt to import some olive trees sadly failed; all died after being detained in a customs shed too long.

The expanding size of the vineyard required us to increase the size of the workforce. After much horse-trading, it was decided that Zhang and Huang would continue to look after the first 100 *mu* on the same terms, whilst the remaining 200 *mu* would be split relatively evenly between Ruan and Meng (both from Qiushandian), Gong (from Mulangou) and Shao (from Xinxing). These negotiations were very delicate, because of the necessity of spreading work to the various villages in accordance with the amount of land rented from them, in addition to the normal personality clashes and differences in ability. Tiffany led the negotiations with a skill which would have made Kofi Anan proud.

As part of an effort to improve the efficiency of the operation we invested in tractors for the four new farmers (they are no use for the first 100 *mu* where the rows are too narrow). This was initially met with deep scepticism (land too steep/too rocky, tractors not powerful enough) but objections were overcome as the tractors were first used to ferry the plants to the fields then, fitted with tanks, to irrigate them with the vast amount of water necessary for young plants. (The tractors were eventually abandoned as impractical on the vineyards slopes and three of the four were sold).

Progress on construction of the castle/winery was frustratingly slow in 2006. The main problem was the need to translate all designs from Ian Begg into Chinese plans consistent with Chinese regulations. Several conference calls, usually early on a Saturday or Sunday morning in the West of Scotland, were held between Penglai Construction's designers and Ian, with us translating in the middle. Certain insuperable Chinese regulations forced major redesign work. The realisation that we might need to cope with busloads of visitors also led to a redesign of the link between castle and winery. The height of the winery was raised to account for machinery, which actually simplified the design for the courtyard above. The sorting shed was shifted from one side of the courtyard to the other as we decided to use an automated method of taking grapes to the courtyard/roof, rather than tractors. Our inability to source a small elevator locally also led to the lift space needing to be increased.

By early 2007, we were in possession of Chinese plans and were able to look for a contractor. None of the contractors we met were terribly convincing; for all of them this project was unlike anything they have ever done before. Not surprisingly, perhaps, quotes all came in above budget and excluding various items, including one quite which excluded all doors and windows! Eventually we decided to work with the government to organise an official tender. The deadline was May 28th. With all the delays, we had to abandon any hope that the winery would be ready for the first harvest, so we decided to make our first wine in the warehouse at the farm. We estimated that we would need five or six stainless tanks to cope with the partial harvest and make about 25 tons of wine. The next project was to find suitable equipment suppliers, preferably in China, both due to cost and to avoid the complexities of the import process.

There is a plan of how our 300 *mu* (21 hectare) vineyard looked by mid-2007, offered as an image insert (I told you it was complicated).

We waited with growing excitement for our first harvest.

CHAPTER 6

First Harvest

Disaster teaches us humility.
St. Anselm

Disaster. That is the only word for it. Everything that could have gone wrong did. Shortly before the 2007 harvest was due to begin, Christophe contacted me demanding a wage increase. When I suggested we could consider it after he had actually made some wine, he resigned. The six 5-ton stainless steel tanks which Christophe had designed with a factory in Tai'an, and for which we had paid a premium price, were delivered late and dumped unceremoniously in the farm yard; the scene of my team manhandling these huge tanks into the warehouse by the sweat of their brow and no little ingenuity would have made a good revolutionary poster. When we came to examine the tanks we found them to be badly designed and manufactured; one of the team was nearly injured when, working inside the tank, the thin wire holding the lid snapped. A key part, the seals around the floating cap needed to keep the tanks airtight were to have been brought from France by Christophe, so we had to bodge up an alternative. Our requests to have the tanks repaired (or taken away in return for our money back) were met with delays and insolence. When we quibbled about paying the remaining 50 percent until problems were rectified, the Taian company froze our Bank of China account, so we were forced to pay or face a court case.

The de-stemmer, press and pumps, ordered from a French company called Fabbri on Christophe's advice, reached Qingdao but then, for reasons never made clear, failed to be unloaded and were taken back to Singapore. Calls to Fabbri were useless; the company neither knew nor cared about the location or likely arrival time of their equipment (already 90 percent paid for by us on shipment). Eventually the equipment, which was contracted to arrive by August 20th, turned up on November 5th, well after the harvest was finished. The equipment sat under cover in the farmyard, still shrink-wrapped and useless for a whole year. Another bad experience linked to doing business with the perfidious French.

Now these were setbacks, but not insuperable. In place of Christophe, we drafted in a Tasmanian winemaker with the good Australian name of Michael Vishacki, who had a Zapata moustache and an old cowboy hat he might have taken off in bed. He made some unusual products from his Tasmanian winery; I will never be able to forget his Sauvignon with Ginseng root. Unfortunately. He did his best with the harvest, albeit expensively and with the odd fit of histrionics ("How is an artiste like me expected to work under these primitive conditions!" I paraphrase). Two of the tanks were adjusted so that they could be used, albeit with difficulty. And Tiffany, Mrs. Lu and friend Mrs. Ren ran around and managed to borrow local equipment to stand in for the errant Gallic equipment. What we couldn't do anything about was the weather.

From late August, it was shocking. Heavy and persistent rain turned the lower part of the vineyard into a quagmire. The grapes sucked up the moisture, the skins split and they rotted on the vine. When we thought things couldn't get worse, and were waiting for the typical dry autumn weather to start, the area was hit by the tail-end of a typhoon, which usually land much further south, flattening several of the trellises. It was a sad sight; piles of rotten grapes had to be buried. I was told that this was the worst weather for ten years, and all Shandong grape growers suffered. All I could do was hope that it was not the start of a new weather pattern resulting from global warming.

In the end, having expected a harvest of 25 tons in early August, we only harvested five tons and ended up with just 14 x 225-litre barrels of frankly unpleasant red wine, together with a large plastic container of even worse. I wondered if perhaps it could be distilled into brandy. (Before the Health & Safety and Tax Bureaus contact me, I would like to say that this production was for non-commercial use only).

So little to show for so much effort. Ironically, the grape which performed best was the one of which least was expected, Arinanoa. This is a Tannat/Cabernet-Sauvignon hybrid which has a particularly thick skin (a virtue, it seems, also useful for winemakers in China). Our Tasmanian winemaker threw up his hands, threatened to give me a Serbian bowtie if I did not pay him immediately, and has never been seen since.

Aside from the disastrous harvest, there were some positives from the year. The survival rate of the vines planted in the spring was high. The design of the new plots, and the system of ditches, saved them from greater storm damage. It was clear that the granite posts in the original 100 *mu* were not buried deeply enough, so we made the posts for the new 200 *mu* longer

(1.7 metres versus 1.6 metres). Some of the wire we bought was substandard (i.e. not properly galvanized) and needed to be replaced. Following our liberal use of chicken shit, we employed a new type of fertilizer at the end of the season – seaweed. In my Monday to Friday job, we have invested in the world's largest seaweed processor, not far away on Shandong's south coast, so we have a sound source of supply.

The government completed its promised concrete road, running it not just from the reservoir to the castle, but on past our farm and through Mulangou, before joining the main road to Qiushandian. Given the steep dirt road it replaces, this is a clear and present benefit to the villagers. A dirt track was also driven half-way up the mountain, giving better access to walkers, tomb visitors and plant gatherers. After some pressure, the authorities also dug us a well, as contracted. Bizarrely the well was drilled about one km away from the castle, down by the lake and suspiciously near a factory owned by the brother of the village chief. A long (and vulnerable?) pipe has been laid to bring water up to a reserve tank at the castle. Before moving away from government matters, I should record a conversation held with one of the farmers:

Farmer Z buttonholed me while on a walk around the vineyard.

"Do you know, Mr. Chris, that the village chief elections are coming up soon?"

"Really?" I replied, feigning mild interest.

"Yes, and I want to be village head," Farmer Z continued.

I probably displayed some surprise and distaste. "Really? I thought it was a Communist thing and the party just picks whoever they want?"

"No!" Farmer Z replied, his feelings hurt. "It's a free election.

The people get to choose who they want."

"Really?" I exclaimed, growing excited with visions of grass-roots democracy flourishing in China.

"Yes!" said Farmer Z with equal enthusiasm. "So if you could just lend me the money so I can buy the votes, that will be fine."

As a postscript to this conversation, Farmer Z was not allowed to stand, being disqualified by having spent a couple of years living in a city. We invited over for dinner the victors in the four villages where we employ labour; one – the new village head for Qiushandian – is a lady and non-communist, which I think impressive in patriarchal and Communist rural Shandong. Later we discovered that her husband, a forklift driver at a local gold mine, was an excellent chef. The others were all good party men, including Shao from Xinxing, the father of our cellar worker, and the local Romeo, "Xiao Shao".

Construction finally got off to a start with a bang, after interminable re-designs and re-negotiations. The impasse on price was broken through a public tender. I attended a meeting in the Penglai government offices when the bids from five construction companies were opened – quite exciting, like an Oscar ceremony: "And the winner of the best construction category is…"

After comments by the panel of experts we decided on the lowest bid (RMB 6.8 million, for the record), which came from the most local firm, Renhe, which also just happened to be the one supported by the Daxindian Party Secretary. The contract, in which the castle is to be finished by the end of June 2008, was duly signed and site clearance was started shortly thereafter. Fireworks, a large, inflatable red arch, dancing and, of course, speeches marked the start of construction, recorded for posterity by the local TV station.

During the summer two representatives from Renhe, Messrs. Wang and Sung, came over to Scotland, accompanied by Ms. Lu, their first ever trip abroad, to look at some real Scottish castles. In the airport carpark, Mr. Wang opened his suitcase to find a coat to protect himself against the Edinburgh summer. Out spilled packets of Chinese snacks, clearly a kind of survival diet, in case foreign food was totally inedible. Our Scottish architect, Ian Begg, showed them around, visiting old and restored castles (including Dairsie) as well as meeting the craftsmen involved. So were the mysteries of lime mortar, sash-and-case windows and leaded lights illuminated. Mr. Sung's comment on being asked his initial impressions of Edinburgh was: "There's a lot of stone." After the initial misapprehensions were overcome (I am not sure whether they consumed all of the peanuts and noodles in their cases) Wang and Sung showed admirable enthusiasm, but as the trip wore they seemed to become gloomier as they realised just how difficult a job they had taken on.

By the time the first snow fell – rather late that year on January 10th – the roof was on the factory and the rafters had been lifted atop the banqueting hall. The speed of construction, as with the farm, was impressive, a function of 30 or so workers living on site and working from dawn to dusk, seven days a week. After a high wind removed the roof of the worker's quarters one night, sleeping accommodation was hastily removed into the basement of the castle which, fortunately, has turned out to be bigger than planned (something to do with ground levels). So far, of course, it has been mostly about pouring concrete, a Chinese speciality; the fiddly bits have yet to come. Nevertheless, with the scaffolding taken down, it's impressive to look down from the viewing platform, above the granite cliff that makes up one wall, into the cavernous space of what will be the winery.

CHAPTER 7
ALL CHANGE

Good judgement comes from experience, and
experience—well, that comes from poor judgement.
A.A. Milne

I haven't failed.
I have just found ten thousand ways that don't work.
Thomas Edison

THE YEAR 2008 was another difficult one (is there any other type of year in China?). This wasn't because of the weather, which could be described as "normal" after the biblical storms and floods of 2007. The problems were down to the difficulty of managing people in China, especially from a distance, as well as the intervention of a malign fate.

The saddest event was the death of old Mr. Liu, who had only been working as the vineyard manager for a few months. He had ironically joined us for health reasons – he was not enjoying working at a high altitude vineyard in Yunnan and wanted to return to his old home. In his short time with us, he was a positive force – always polite, despite the difficulties of persuading the farmers what to do and when to do it.

In my memory, he always has a smile on his tanned walnut of a face. He died in a sudden summer shower in the vineyard when, hurrying back to the farm, his motorbike ran into a truck carrying sand dredged from the lake. The accident took place in front of manager Feng; an ambulance was called but Liu was dead on arrival at hospital. Unfortunately, we then discovered that Feng had failed to pay the staff insurance premiums, so that the company had to bear the full cost of the pay-out to his widow. Liu, like village head Huang before him, was not wearing a helmet, but there is no sign of these salutary lessons having any effect on the village's remaining motorbike drivers. The over-loaded sand trucks were temporarily stopped from using our road by a height bar we constructed across it, but this was soon dismantled by the same truck drivers.

We found a new winemaker in Mark Davidson, owner of Tamburlaine Vineyards, whom we had met during our holiday in the Hunter Valley in New South Wales during Chinese New Year 2007. I had been impressed by the quality and consistency of his wines, and am sympathetic to his organic approach to growing grapes (a strange and deviant heresy from the point of view of China's chemical-intensive farmers). He already sells his wine into China, and is an investor in a wine bar in Guangzhou, so has reasons to head to the Northern hemisphere regularly, when this does not clash with the harvest in Australia. This year Mark's main contribution has been in the design and installation of the winery equipment, which meant that we have adopted a New World low-temperature approach, rather than the more traditional French approach which has so far influenced most Chinese producers.

On one visit to the vineyard I met a Mr. Hao, who had studied wine-making in Germany, before returning to work in a lychee liquor-making company in Guangzhou. He was looking for a job in wine closer to his home town of Pingdu, Shandong. He seemed a practical and sensible young man, and the foreign experience was a definite plus, so I hired him to manage the winery under Mark's guidance. Hao in turn recommended a classmate from Pingdu Agricultural College, a large individual called Kang, with experience in vineyard management with Huadong, and a number of other local wine companies.

Our longest-serving employee, Mrs. Lu, had long been critical of Feng's management skills and lack of specialist knowledge. Feng's expensive mistake over the insurance premiums allowed her to intensify the campaign. Junior staff, Edward and Cindy, complained of Feng's autocratic behaviour. It was intimated that all staff would depart if Feng was retained, and Kang made it clear that he would not join as long as Feng was in place. We therefore decided to fire Feng and appoint Hao as manager to replace him. This is not what Mrs. Lu wanted at all. She immediately adopted a policy of passive resistance towards Hao. She neglected her responsibilities – construction of the castle slowed to a crawl. Rumours that she was taking back-handers on our purchases surfaced (the case of the rusting "galvanised wire" was raised). These were never proven, but given some validity by the discovery that she had bought several properties, including her own 30 *mu* vineyard. Once she declared that she was unable to co-operate with Hao, she had to go, her husband bank manager notwithstanding.

The genial cook, Mrs. Fang, her relative, followed shortly thereafter. We never saw Mrs. Lu's perm again, but Feng became the manager of a new neighbouring vineyard, Xiandao.

This purge, of which Mao in his heyday would have been

proud, meant that we finished the year with a rather different team from when we started it.

Another person missing was Mrs. Zhao, our nice cook from Shanghai, who we had been training in Western cooking with the future castle restaurant in mind. She had been working for us for a year, living in our house in Shanghai to escape her scoundrel of a husband, who used to beat her. When she asked for a loan of RMB 20,000 to allow her chronically unemployed husband to offer a taxi service, and start to provide for their two daughters, we were delighted to help. She would gradually pay us back out of her salary. Yes, you guessed it. After we handed over the money, we never saw her again. The addition not yet mentioned is the pretty interpreter Eliza Zeng, who became a full team member at Mark's urging.

The turnover of staff was not conducive to any big improvement in management of the vines. We eventually completed installing granite posts and wire on the 200 *mu* planted in 2006, though this was, bizarrely, delayed by the Olympics.

Scared of terrorism at the Beijing games, the authorities restricted the ability to buy explosives, which meant we could not get powder to blow holes where rocks prevented the insertion of posts.

The construction company used a related excuse for the slow-down in construction at the castle; apparently quarries were forced to curtail the supply of stone.

Old Liu's plan to allow native grass to grow between the vines was not entirely successful. The farmers of the original 100 *mu* simply would not co-operate, so continued to weed the ground between the vines. The farmers of the remaining 200 *mu* complained bitterly about the time required to cut the grass. Next year, we decided, we would try clover, which keeps low and has the added advantage of staying green longer into the winter. One achievement was a stone-clad chemical mixing tower; this allowed us to control the mix of the chemical sprays supplied to the farmers, which otherwise tended to be applied to their own crops. Once we moved to direct labour, later in the story, the tower was no longer necessary. So it now stands disused, a strange memorial to Old Liu

Despite the moderate weather, we still faced problems of grapes rotting on the vine which, together with peer pressure to get the harvest in, meant that the grapes were picked too early. The whites were harvested from August 23rd. All the reds were in by mid-September. The actual amounts harvested are shown below.

Production of all grape varieties in 2008

Varieties	Mu Quantity (mu)	Total Production (kg)	Per Mu (kg)
Muscat	2.28	963	422
Viognier	17.7	4914	278
Cot	10.38	1901	183
Merlot	12.06	1759	146
Syrah	11.43	2258	198
Cabernet Sauvignon	10.12	937	93
Sangiovese	8.53	4989	585
Cabernet Franc	12.15	2870	236
Petit Verdot	9.6	2493	260
Arinanoa	10.03	2230	222
Total Amount	104.28	25314	243

Resolutions to improve vineyard management in 2009 under the new Hao/Kang/Davidson regime were as follows:

- use the winter to improve drainage in some of the damper parts of the vineyard
- raise the first wire to 90 cm from the ground
- develop a more effective spraying plan
- reduce yield through a one bunch per vine policy
- move to a system of direct labour

The last of these will be the biggest change. The problem so far has been that the farmers have their own land, which naturally takes priority over the vineyard when any urgent activity is required. The cost of direct labour will be higher – in addition to salaries we will need to provide board, lodging and all equipment – but we will be able to instruct them to do what we want, when we want it. Previously much of the managers' time seemed to be spent in a.) finding the farmers, b.) trying to persuade them, not always successfully, to do the necessary and c.) making sure that our materials – fertilizers, chemicals, posts, wire etc. were only applied to our own land.

We introduced the direct labour scheme when the existing contracts expired in March. The other main obstacle to producing a decent bottle of wine from the 2008 harvest was that the winery was not ready – a function of Mrs. Lu's go-slow campaign and the turnover in personnel. We had plenty of stainless steel tanks – in fact too many. In addition to the six 5-ton floating-cap tanks from last year, eight huge double-layer tanks (10 ton fermentation/10 ton storage) were constructed on-site. This overcapacity (80 tons fermentation and 110 tons storage, compared to a harvest of just 25 tons) proved ill-timed, given the subsequent slump in stainless steel prices. The new refrigeration system was not connected. The roof-top sorting system was not completed, which resulted in more mouldy berries entering the tanks that a more careful sort might have eliminated. We had to use a diatomite filter, rather than cross-flow filter. The waste system was not built, leading to an embarrassing stream of grape skins across the road and down the bank in front of the castle as waste systems in China are viewed as a luxury, not a necessity. The lack of some small accessories, such as a stainless slide to link the fermentation tanks to the top of the press, made work unnecessarily difficult (workers balancing on top of the press, pulling skins from the fermentation tank down a slide made from a large plastic bag). The laboratory was not installed, making tests unreliable. And all around us, dust raised by the continuing construction of the castle swirled.

Given all this, the small amount of white we produced – a Viognier/Muscat mix – was okay, exhibiting some floral character (I wouldn't send it back if served it in a restaurant). The reds and, especially, the rose, were thin with a hint of mould (these I would send back). Taking some samples to Great Wall's huge Junding complex, and tasting their 2008 wines, it was a consolation, albeit a rather a sour one, that with all their money, their reds were also thin and mouldy. Is it possible to produce a decent red wine in Shandong? Should we have concentrated on early-ripening varieties and planted more white?

We bought a mixture of 300-litre French and American oak barrels from a local Australian supplier. We also tipped out the worthless product of our 2007 harvest into a plastic container and reconditioned those barrels. All were now filled with our 2008 product and laid in the warehouse at the farm, which has a warmer environment to help with malolactic fermentation. We were to buy a bottling line in 2008, but the prospect of "a decent bottle of wine" remained distant.

We had our first paying guests at the farmhouse. My university friend Lindsay, accompanied by her unpleasant boyfriend from Hong Kong, visited Mulangou. Apparently the boyfriend had thought he was spending the weekend amidst the bright lights of Shanghai, rather than stuck in rural Shandong, which probably accounts for his ill-humour and heavy drinking. We also had a visit from the better part of the board of China Fund Inc., the New York listed investment fund which I was managing at this time, ahead of a board meeting in Qingdao. Early in the year, we even had a Scottish politician visit; Penglai is twinned with Angus (?!). It all makes for good practice.

Construction of the castle, due to be completed under contract by June 2008, then in October, then in November, dragged on. The construction company was under pressure from rising material costs, as well as the unexpected time required for stone work, with which they have little experience. As this is a fixed price contract, this shouldn't be our concern. Unfortunately, out of the goodness of our hearts, and no little stupidity, we had paid ahead of the building schedule, which meant there was a real danger that the construction company would simply walk away, leaving me with an unfinished shell. This outcome was temporarily avoided through a mixture of jollying along and pressure from our political friends (the construction company derives the larger part of its business from government contracts). Compromises were made, as normally happens in any building project; most concerned the size and thickness of the granite used, so shouldn't be evident to the layman, but it would have been nice to have the arched brick roof in the factory. We also lost the solar heating beneath the courtyard when it became clear that the builder's did not have a clue what the architect was trying to do. We also made some changes to the layout, sacrificing a dining room to allow a full complement of six good-sized bedrooms.

The next task is interior decoration, which I rather dread. We will not leave this task to the construction company after having seen a builder's idea of finishing at the farmhouse, but it will be difficult to find a decorator with any conception of what we are trying to achieve. New luxury projects in China are inevitably shiny, with lots of polished marble, redwood furniture and golden chandeliers – preferably with a fountain.

The 2008 prize for daftest construction went to the suppliers of the grape elevator, the job of which is to lift the baskets of grapes onto the roof of the factory for sorting. Apart from its technical flaws (lack of weather-proofing) the company decided to use blue glass. This eye-sore galls me every time I visit the site.

As the prospect approached of having something to sell, we spent some time on packaging and marketing. We completed the website (www.treatyport.com), adopting an old, sepia-photograph style for the labels which harks back to the time of the establishment of the treaty ports in China. The website remains rather clunky to update, and will need to be modified once we have more experience of trading online and have finalised our product line-up and membership plan. On one day, back in Yorkshire, I proudly entered the website address, to show my parents what we had been up to, only to find a blank page telling me that the website had been hacked, in Turkish.

The Edinburgh-based design company, Size Design, worked up some branding/packaging ideas for me. This sparked much debate, but opinions split along cultural lines: nearly all foreigners preferred the old label-style packaging; nearly all Chinese preferred the label with the 16th century gentleman (a portrait by Gheeraerts the Younger which will hang in the completed castle). The thinking of the foreigners is that the label idea is unique and links back directly to the treaty port heritage. The thinking of the Chinese is that foreign wine is better than Chinese wine, so the more foreign the packaging can look the better.

Size Design then went away to try to come up with a middle way, with old sepia photos of foreigners in China. This pleased no one (apart from me); the colonial characters were felt by both sides to be too scruffy to represent a premium product. Perhaps we should look for photos of more elevated foreigners? The search continued.

There were plenty of "political" lunches in 2008. On every visit to Shandong we met with Daxindian Party Secretary Zhong, usually inviting him over to the farmhouse for lunch. We ate with our most senior friend, the large-faced Yantai Party Secretary Liu – usually accompanied by his old friend, Mrs. Ren, whom he had recommended to us as an advisor. I helped Liu's daughter, Sarah, a bright girl and recent LSE graduate, by employing her as an analyst in Shanghai for most of the year before I recommended her to Taiwan's largest broker, Yuan Ta. We also met up regularly with Penglai Mayor Zhang, who had given the speech at the start of castle construction and claims to want to retire there. To each of the meals I took along a couple of bottles of Tamburlaine wine, partly as propaganda, and partly to prevent me having to drink the horrible stuff they serve otherwise (local grape wine or, even worse, local *baijiu*). Party Secretary Liu was an honourable exception, on one occasion bringing along a lovely Bordeaux wine from a vineyard recently bought by a Shandong businessman (the brand was called "Ney Michel"). At another dinner, hosted by the Australian barrel supplier (and new inamorata of our translator, Eliza) a Taiwanese friend kindly bought a bottle of vintage Chateau Palmer; unfortunately it was badly corked, but the table continued to drink it with every evidence of pleasure, even after winemaker Mark had pointed it out.

Our main political objective, apart from *guanxi* maintenance, was to keep pressure on the construction company to complete its contract. We finally got the deeds (*"fangchanzheng"*) for the farmhouse, but were still pressing for the castle deeds (which we did not receive until 2014). We mooted the idea of buying another plot to build another *"siheyuan"* courtyard-style farmhouse, to extend our accommodation, and build a cellar for storage. However, I was warned off the spot I chose, which had a good view and no cultivation, as it was "unlucky"; on further enquiry I was told that two sisters had hung themselves there. We still did not have a license to sell wine (this requires ownership of a bottling line, a lab and a sign off from the food hygiene bureau), but received permission to produce wine "for our own consumption".

Towards the end of the year, a U.S.-educated private equity specialist from Taiwan, K.C. Kung, who had heard of our project through our mutual investment in a local pharmaceutical company, decided to plant a 100 *mu* vineyard in our valley. With our blessing, he hired our ex-manager Feng to help him. He chose a nice spot on a peninsula overlooking the lake. So far so good, from my point of view, and I told him I was happy to share the fruits of our hard-earned experience. The more vineyards in the area, the more of a destination it will become. Unfortunately, the environment bureau told him that he would not get any deeds on a factory built within 500m of the shoreline. Over time I am sure ways and means could have been found to get the deeds, especially if the initial plan was just to put up a restaurant/ guesthouse. But the gentleman had spent too much time in the U.S., where the law is applied rather differently. So he announced he "had no alternative" but to put his house up on the slope of Qiushan, above our vineyard and alongside the castle. At this point he lost my support. He became abusive when I told him I would do my best to prevent it. (He later calmed down and is building a winery above a quarry about a mile south of the castle).

A new experiment in 2008 – unlikely to be repeated soon – was the offering of a summer holiday job to one student of Chinese at Oxford University. I had enjoyed my time studying Chinese at Oxford, so in the spirit of giving something back, I contacted the university. As my eldest daughter Matilda was also at Oxford, and was in charge of interviewing, it was not perhaps entirely surprising that the chosen scholar was a tall, handsome young man called Michael Wormer-Oates (he can be seen grinning in a photograph of the visiting China Fund Inc. team). The experiment was not an unqualified success, partly because of all the turmoil going on in such a young company, partly because of the natural deferment of the Chinese staff to a foreigner, which meant that Michael was not required to do much work. Still, this did not show in his end-of-trip report, which I publish in full here to show you the continuing value of an Oxford education in being able to put a positive gloss on almost any situation:

> I stepped off the train at Yantai and looked around. I was in a part of China I'd never been to before, about to work with a group of people I'd never met doing something I'd never done before. As I filed off the platform with all

the other weary travellers, I began to think that a less adventurous summer of clubbing and sunbathing might have been a wiser choice. That thought disappeared within five minutes of meeting Mrs Lu (the site manager for Treaty Port) and Edward Tian (head of marketing). On that sunny Tuesday and for the next two months, they and everyone else at Treaty Port would be welcoming to the point where I felt more like a friend than a colleague.

I had no idea of what to expect from Mulangou. It is on no maps (no, not even Google Earth) and, as far as I could tell, surrounded by miles and miles of fields. I had visions of spending my summer in isolation with only cows to talk to and porridge to eat. This could not have been further from the truth.

On the way to the village itself, I spent more and more time gawking at the countryside out of the taxi windows. Far from being desolate, Shandong is pretty much the orchard of China, with every possible piece of land being crammed with fruit trees. On a sunny day (and there were many) the sight and smell of the fruit ripening on thousands of trees was amazing. Mulangou itself is a small, friendly village of roughly 500 people which sits in the shadow of Qiushan and enjoys views over rolling orchards and the, spectacularly blue, Qiushan lake. It's the kind of place that, if it were in Italy or France, would grace the cover of every travel supplement and be mobbed with Brits on holiday. Thank God it isn't!

Treaty Port owns a restored farmhouse in Mulangou village. Every one of its 6, whitewashed bedrooms were eerily reminiscent of the ones I had left behind in St Anne's. They are spacious and comfy and, most importantly for keeping up with the rest of the world, all have internet connections. They are a great place to avoid the heat of the Shandong midday sun and just as good for relaxing at night.

Things looked pretty good so far but get even better at meal times. If you don't like Chinese food, don't let this stop you coming to Mulangou. Mrs Fang (the cook) will convert you to the delights of 馒头 so quickly that you will, like me, have to start jogging every day to keep the weight off! Meal times were a great chance for the eight main employees of Treaty Port to get together and banter about everything under the sun. Any nerves you might have about being thrown into a fairly alien environment will soon disappear under the barrage of banter than will come your way at breakfast, lunch and dinner.

While the combination of manual and more white-collar work (I, for example, helped crush the 2008 vintage's grapes and then headed home to translate the company website into English) will keep you both interested and on your toes, the best part of working at Treaty Port is the chance to meet new people, discover new things and make new friends. Within a week of arriving at Treaty Port, I felt more like I was joining a family than a company. Going out to dinner with colleagues was always a treat but left me feeling slightly embarrassed since the village's one shop didn't really sell any decent presents for my hosts! Living and working at Treaty Port won't just improve your Chinese, you will come home knowing more about making wine than (going by the taste of their wines) any of Treaty Port's competitors.

Not only did I have the chance to spend a great two months with the guys at Treaty Port, but the constant through-flow of visitors kept life in a sleepy village from dragging.

Mid-autumn festival with local cadres, a beer (or two) with Neville and Mark and lunch with Yantai's very own king of stainless steel, Mr Xiang (if you don't meet this man, your trip isn't complete), these are just a few of the unforgettable experiences I had.

When I got back on the Beijing-bound train at Yantai, I felt genuine sadness at leaving the people who had been such a large, friendly part of my life for two months.

Treaty Port is not an average 9-to-5 company where people work in cubicles and leave promptly; thanking God another day is over. It is more than that. It is a place full of great people who, while they care deeply about their

```
jobs, never let this stop them from enjoying
all the other things in life. While working at
Treaty Port, I saw the company grow and felt
like I was really making a difference. This was
certainly the most fun, rewarding, challenging
job I have ever had.
```

I will also complete this chapter on an upbeat note. It had long been rumoured that Lafite was looking at establishing a vineyard in the area. The authorities had been careful to make sure we never met, on the grounds that I might not give an entirely glowing picture of my experiences to date. They need not have worried, of course, as having Lafite as a neighbour could do nothing but boost the value of my enterprise (or "rural charity," as Tiffany had started to call it). So I was delighted in December when Christophe Salin of Lafite visited me at my Shanghai office to announce that the negotiations had been completed. In a 70/30 venture with a large Beijing-based conglomerate, CITIC, Lafite plans a 25ha vineyard, its first in Asia, which will be doubled in size if all goes according to plan. There was additionally a larger "exclusion area" over which Lafite has an option, to protect the scenery from unsuitable investments (which strangely seemed to include our vineyard). The company planned to spend a full year testing the site, which lies just over the hill to the South of our vineyard and will not plant vines until 2010. Mr. Salin was anxious to hear my views on Gerard Colin, whom they have hired as a consultant. I was suitably diplomatic.

At a subsequent meeting with Christophe Salin, I remember standing together on a balcony at the castle, looking out over the verdant summer countryside.

"What are your plans for the winery? I expect your partner,

CITIC, would like a nice Chateau with visitor centre?"

Christophe paused, gave a gallic shrug and the said (you will have to imagine the accent): "We have no such plans. We are simple farmers."

"Yeah, Right!" Actually I did not say that. I am much too polite.

CHAPTER 8

IT'S TOO LATE TO BE PESSIMISTIC

There is no mistake as great as the mistake of not going on.
William Blake

If at first you don't succeed, try, try again. Then give up.
No use being a damned fool about it.
W.C. Fields

THE BACKGROUND TO this year's activities was the global financial crisis and economic recession. This did not have any direct bearing on the project, but added a certain piquancy to my situation, as cash continued to flow out and we stumbled from one crisis to the next. The nadir was probably reached in January, as stock markets plunged, when both the water and electricity ceased working during our stay at the farmhouse, and the thermometer inside our bedroom showed -4C°. I remember Dr. Guo once remarking that he was impressed that I wore a clean shirt every day. Ice forming on the inside of the bathroom window is certainly a disincentive from prolonged ablutions.

When the construction company, Renhe, downed tools for the long winter break, one corner of the castle roof was left deliberately untiled. This was to make it less likely that we would fire them. Meanwhile their two idle cranes continued to loom over the site, warning off any other construction companies that might be tempted to finish off the project. The builders continued to wage a guerrilla struggle throughout the year.

As we needed them to sign off the completion document for the castle, in order to get the deeds, I bit my tongue and we continued to try to jolly them along, even as further evidence of their incompetence emerged. We relieved them of all sensitive work (windows, doors, lift and other interior equipment) but gave them work on the outbuildings and approach road, to help their cash flow. We used our political contacts to bring as much pressure to bear as possible, including a tough lady official from Yantai in charge of government construction.

"Well thank you so much for your help, Mrs. X," I said somewhat unctuously. "We really appreciate it. If there is ever anything we can ever do for you, you have only to ask."

"Handbags," she said, setting aside her clipboard.

"Handbags?"

"Yes, Handbags."

Apparently the key is to leave the original invoices inside, to show that they are real brand-name handbags, and not fakes. Mrs. X subsequently told me that it was all my own fault for hiring such a bunch of numbskulls in the first place. My objection that they had been hired through a government-organised tender process, overseen by "experts" was disdainfully ignored as irrelevant. The final bill was not yet in, but despite their many omissions and mistakes, it became clear that the construction cost would be substantially more than agreed.

After our bad experience with the builders, we approached the hiring of an interior decoration company with trepidation. After detailed briefings from me, using my well-thumbed books about Scottish castles and Jacobean design, the resulting proposals mostly resembled theme park rides or the decor for a new fast-food chain.

We were delighted, therefore, to be introduced by a Taiwanese property developer friend, Li Ziqiang, to a Taiwanese decoration company with long experience in China.

A delegation came over for dinner to our Shanghai home. The owner, C.K. Lin, was reassuringly bearded. His sister-in-law and designer, Lillian, had actually studied in Scotland. Their construction expert, Li Gong, conveyed the right balance of cynicism, experience and hope. In his first site visit, he stood up to the construction company, Renhe, and clearly knew what he was about. This Taiwanese company would be more expensive than the locals, and also would not produce any designs until money was paid over, but we decided to employ them.

This did not prove a successful decision either. It is not that choosing one of the local decorators would have been better; problems would probably have been papered over to emerge later and the low bids would have proved as mythical as Renhe's. But C. K. Lin did not live up to his billing. Despite the stay in Scotland, the designer showed no more sense of what a castle interior might look like than the locals. Worse, Li Gong turned out not to be part of C.K. Lin's company, but to be a subcontractor whom the company seemed reluctant to pay. Other commitments prevented him from being based in Shandong, instead sending a well-meaning but ineffectual friend, Mr. Gao, who scraped his few grey hairs into a ponytail, and was eventually fired.

The detailed budget and timetable promised at the start of the project, never materialised, as Li found that his experience in the Shanghai area had not prepared him for the greater difficulties of operating in the unfamiliar Shandong countryside.

In the end, he hired Bangtai, one of the local decorators which had previously quoted directly. This all sounds terribly negative, but several problems were caught: Renhe's decrepit underfloor heating system was dug up and replaced; the expensive wood windows, initially installed so that rain ran down the inside, were properly re-installed; the pipes for waste water and sewage were unlinked; a drainage ditch was dug against the North wall to prevent rainwater running into the basement. The final bill remained to be tallied, but it looked as though it would not be too far from the original RMB 2 million budget.

In the vineyard there were some significant changes. We moved to using contract labour, directly hiring a team of ten to twenty labourers on temporary contracts, providing accommodation in cottages rented in Mulangou. The labourers were split into teams under Mr. Zhang (our original farmer from Qiushandian) and another Mr. Huang, both under the direction of Kang. The average age of the workforce seems to be well over 50, but it perhaps makes up in steadiness what it lacks in speed.

The whole vineyard was left to natural grass cover. This resulted in a rather messy appearance until we bought a large mower late in the season (we needed to buy a second for 2010). But it is effective in preventing soil erosion and should gradually build up the organic content of the soil. The bottom wires were not raised to hip height, as has been required by Mark to help air circulation and counter mildew problems (to be corrected in winter 2009/2010). We also failed to fertilize this year; Hao messed around finding the right material until it was too late to apply. I decided to limit crop yield, requesting we retain just one bunch of grapes per vine. This was not fully applied, as the workers regard such green pruning as throwing away perfectly good grapes. However, the lack of fertilizer and overuse of hydrogen peroxide as an anti-mould drying agent resulted in a crop, the first from the whole 300 *mu*, of 38 tons. This compared with an early August estimate of 80 tons. Still, good weather, combined with the reduction in yield, resulted in our best harvest yet, with no grapes with a sugar content less than 21 percent, and some grape types as high as 30 percent. This was substantially higher than anything achieved by the other local vineyards, so resulting in some admiring visitors looking in to confirm our claims and find out how it was achieved. It was noticeable that better results were achieved from plots higher up the hill, and therefore with better drainage and exposure to the drying breeze. We started the harvest on September 3rd with the Merlot, and picked the last Anna on September 21st.

2009 harvest by grape type

Varieties	Production (2008) Kg	Production (2009) Kg	Brix (2008)	Brix (2009)
Cabernet Sauvignon	937	4,349	19.5	26.10
Viognier	4,914	3,713	14.4	21.96
Chardonnay		620		21.96
Petit verdot	2,493	841	16.5	30.6
Cabernet Franc	2,870	1,398	18	30.6
Cot	1,901	1,823	18.5	25.2
Arinarnoa	2,230	800	18	30.6
Sangiovese	4,989	1,910.5	18.5	25.2
Muscat	963	513	14.5	22.68
Syrah	2,258	4,737	17	22.86
Grenache noir		3,192		26.64
Marselan		6,157		25.11
Melot	1,759	7,773.5	18.5	21.06
Total	25,314	37,827 .	average 17.34	average 25.43

Despite my best endeavours, and many meetings with Hao to review lead times, the factory was still not ready for the 2009 harvest. The sorting was not done on the roof as originally planned (the pipe was not connected) but grapes, together with dust and flies, came in through the main factory door, which was not fitted until after the harvest was long completed. The sorting table itself was immediately abandoned as being noisy and impractical. Stainless slides to ease getting grape skins from the tanks into the press were still absent. Ironically the small and varied nature of the harvest meant that the 14 large double-layer stainless tanks which we had constructed at great cost were little used. Instead more use was made of the original small, variable capacity tanks (still without proper seals). The only saving grace was that the global financial crisis had caused a collapse in the price of nickel, and therefore stainless steel, reducing each tank's cost compared to 2008 (RMB 90,000 per 20t tank, compared to RMB 118,000). Renhe Construction had not got the factory floor slope correct, so, despite the stainless steel channels, a large puddle formed in the centre of the winery, through which employees splashed. Furthermore the waste treatment plant was not completed on time, so our neighbour's pond once more became a wine-dark sea.

The bottling line and laboratory were also not ready. This is not a direct hindrance to making decent wine, of course, but in China it means that you cannot gain QS certification. So our wine was once more produced for personal consumption only...

Working in these far-from-ideal conditions, Mark Frazer, who had been drafted in from Mark Davidson's Australian winery, co-operated with the home team of Shao and Adele, and did well to produce what they did. We bought another twenty-eight

300-litre hogsheads from Alan, and re-used our existing stock of barrels, cleaned after emptying 2008's unsatisfactory vintage into the stainless tanks.

When I decided two years previously that the castle's opening date would be October 11th, 2009, this seemed like an easy task. Renhe would be finished by June 2008 and interior decoration three months thereafter. This would leave us plenty of time to sort out any teething problems before the directors of five of the investment funds I was managing, plus management from my Scottish JV partner Martin Currie and local politicos all turned up. Wrong. There were several occasions in the run-up that we were on the point of shifting all accommodation, dining and meetings to a hotel in Penglai (we had one on standby). Even the week before, the castle looked like a disaster area with an army of cleaners making no headway against the mess created by builders and decorators. Work seemed to be going on in every room. Including the landscaping and the farmworkers, I calculated that I was employing about 200 people in the first week of October 2009. I expect there is a blip in the historic records of Shandong's GDP attributable entirely to the Scottish Castle.

My visions of calmly walking round the castle and hanging pictures (the fun bit) were far off the mark. The four-poster bed that the first guest was due to sleep in on Saturday night was still being assembled on Saturday morning. The buzz-saw cutting shutters for the windows could be heard in the courtyard. The lift was never delivered (locked doors were installed over the lift doors so that no guest fell down the elevator shaft).

Some doors, gates and ironwork were missing. The fireplaces, acquired from an English company in Yantai, were not properly installed. The day before the event, a pine forest and verdant lawn suddenly sprang up along the approach road to the castle where only builders' rubble had been. Miraculously the castle was ready for its first guest. For the record, the first guest was Sir Tim Kimber, the chairman of Taiwan Opportunities Fund, the first fund I established in 1995. The Chinese officials were terribly excited to have a real live English aristocrat present (my explanation of the honours system had been partial). Tim was given prime position in the ribbon cutting the next day, and, tall and imposing, with impressive eyebrows, he certainly played his part well. The Duke of Edinburgh couldn't have done any better.

The day of the opening was a continuation of the beautiful autumn weather we had been enjoying, bright and blue. It was pleasantly warm with a light breeze fluttering the flag with its drinking dragon, raised for the first time on the flagstaff above the tower. The vineyard looked impressive, with the leaves of some of the earlier varieties starting to turn red and yellow. A light haze lay over the lake, but the three-fingered peak of Aishan National Park was still visible in the far distance. The ponds dotted around the orchards were low, reflecting the recent lack of rainfall. In a break in the action, I walked through the highest plot of Grenache with my daughters, gleaning the few grapes left over by the pickers, sweet and warm to the taste.

We had a good turnout for the opening ceremony, which started at exactly 5:18 (the number sounds auspicious in Chinese). Most of the politicians with whom we had had dealings were present, with purple orchids pinned to their lapels. The directors and Martin Currie staff were suitably be-tartaned. Ian Begg was there in his kilt, as was Mark Davidson, though his outfit clearly dated from an earlier and slimmer vintage.

The staff also looked smart in their tartan uniforms. Tiffany and myself made a grand entrance, arriving in a Treaty Port–

blue London taxi cab. This was a recent purchase, assembled outside Shanghai, now that Manganese Bronze has been bought by local auto maker Geely. The company driver had taken a day to drive it up to Shandong, without plates, only being fined RMB 600 by the police along the way (no receipts were given). We were dressed in Jacobean costumes made for us by the wardrobe department at the York Royal Theatre, mine patterned on the clothes worn by Sir Henry Wotton in the 1620 portrait now hanging in the Great Hall. This caused something of a stir and was I hope, given the discomfort, memorable. The speeches were not memorable; speeches by Chinese politicians rarely are, tending to favour the formulaic to any real expression of feeling, the dull process lengthened into greater tedium by translation. During my speech, Eliza, dressed in full princess regalia, made an unscheduled visit to the stage, politely greeting the guests and bowing to warm applause. When the speeches were over, she helped me cut the opening ribbon, brought on by tall ladies in red *qipao* dresses (another indispensable item in Chinese opening ceremonies). Then, with the sun setting, the fireworks started.

The guests were ushered through the shop to the Great Hall, with gratifying expressions of surprise and wonder. It looked splendid with the long table laid for 60.

Jamie Skinner, a colleague working for Martin Currie, greeted the guests with a brief but enthusiastic skirl of the bagpipes from the minstrel's gallery. Tiffany, helped by Mrs. Li, Cai Hong and an army of enthusiastic students from Yantai University, engineered a multi-course Western meal. Mark, with some judicious blending, served our own wine, both white (a blend of the 2008 and 2009 Viognier, Muscat and Chardonnay) and red (the 2009 Cabernet Sauvignon and Marselan). This seemed to go down pretty well and a large quantity was consumed.

Apart from bagpipes, the entertainment included a violinist, a far-too-loud rock band, talks from Mayor Zhang and Mark, and renditions of Frere Jacques and Incy-Wincy Spider by Eliza. The Chinese politicians gradually slipped away during the meal (as is their custom) but that did not seem to reduce the noise level. On the overhead screen we showed photos from the setting up of Treaty Ports. The next few days of activities seemed to pass off well, with all guest rooms full, breakfasts and lunches served in the Great Hall or library and fund board meetings taking place in the library.

At this point, I need to include the roll of honour for services beyond the call of duty. Our Taiwanese friends Li Ziqiang and his wife Judy made a big contribution in pushing the project along so that the castle was habitable by D-Day and our management system organised. Tiffany made an incredible effort to feed sixty guests, in style, with limited facilities (using mostly the Aga in the family kitchen, rather than the commercial kitchen). In the process she lost 5kg! I also lost some weight, ending up the opening day with a temperature of 39C° on an intravenous drip, inserted by a local doctor with dirty fingernails. Apparently the cause was an infected insect bite. I had thought it was just the Jacobean costume which was hot.

Another couple whom I must thank is Roman and Ewe Flezar from Poland. They are responsible for the murals which can be seen around the castle, painted in difficult working conditions with great good humour. My abiding memory of Roman is his descriptions of wine, drunk together over dinner, done entirely with his hands, inscribing taste graphs in the air. He was also the main contributor in determining the correction proportion of whisky to ginger wine in our Whisky Mc (3:1, as it turned out, after extensive experimentation).

In the main staircase mural Roman used real faces for the characters. In the portion shown above the nobles have the faces of our winemakers, in historical order, from left to right. I am the monk.

On a less positive note, Mrs. Ren, who had been a helpful if volatile force in the project's development, turned difficult. The catalyst was our decision to grant her a RMB 400,000 loan to help her buy a house at a local spa. We thought that this gesture, together with the expensive tea which we had bought from her teahouse for "presents" to officials, was a reasonable recompense for her assistance. But she suddenly started demanding large sums of money in a barrage of short messages and long phone calls.

At her request, both she and her husband, alongside the mayor and party secretary, had been painted into Roman's mural. Now I felt like painting it out, in the same way that the traitor Lin Biao was expunged from photographs with Mao. But the faces remain, a testament to our gullibility. We never saw Ren again and our RMB 400,000 loan is a "receivable".

One character who painted himself out, so to speak, was a Chinese-speaking American broker called Bruce Richardson. He was "between jobs" so would, I thought, be a helpful presence in Mulangou in the push to get everything ready for the opening. He survived about a week before disappearing one morning, without telling anybody, sending me a subsequent e-mail concluding that my project was hopeless.

Other minor irritations included a hillside fire, perhaps set off by burnt offerings made at a grave, which destroyed some trees and caused damage to piping. There was also the theft of a PC owned by a student worker from the dormitory during the opening ceremony. This crime was never satisfactorily solved.

We have kept going despite these local difficulties; as one economic forecaster said at the start of the year, "It's far too late to be pessimistic".

In the spring, Lafite, which was testing soil in their proposed vineyard, asked if they could dig some sample pits on our land, on the basis that we would get to see the test results. I only received them in December, after much prompting. The report was particularly critical of the prospects for our second planting (though this might be because that part was organised by Christoph Koch, the successor to the Lafite's new advisor Gerard Colin.). I attach an extract:

Synthesis of my observations after the study of the vineyard TREATY PORT. All of pits opened show the homogeneity of the geological substrata, we find the same granite features. We can characterize these soil types as: shallow sandy soil on reasonably resistant granite.

The period of planting in 2007 in my eyes seems to have been unsuccessful, both technically (with the use of the Ripper) and with the choice of manure and vegetation.

The soil was too shallow to have vines planted on it. There is virtually no earth.

Apart from the Grenache 110R which I feel is a big mistake, we can see that the Rootstocks used (Gravesac, 3309C, 101-14) belong to a type of rootstock which are not particularly vigorous. Even though they are not very vigorous, in the Penglai context (geopedology and climate) the vigor of grape varieties are still slightly

```
too high. I would advise you to try to use
grass to restrict the vigor and the production
potentiality, but also to improve the winter
pruning to control the vigor and the production
(numbers of clusters of grapes).
```

Oh well. Gerard, who has restored a house in Mulangou village, attended both the opening ceremony and an earlier event, the whisky dinner. We started importing eight-year old single malt whisky to sell alongside our wine (what is a Scottish Castle without whisky, after all?) and organised a tasting dinner to coincide with an Asian tour of Jonathan Scott, a representative from our supplier, Macleods. I had imagined this as a warm-up event for the opening, but by September 20th, the castle was no way near ready, so the event was held in the courtyard of the farmhouse. About forty people attended, to learn the difference between Lowland, Highland, Speyside, Island and Islay. Our mail shot having born little results, attendees were mostly the usual suspects, with the exception of two Englishmen working in Qingdao, both called John, with their much younger Chinese girlfriends, who giggled a lot and seemed particularly keen to learn what a Scotsman wore under his kilt. I had been persuaded, given the event, to wear a kilt for the first time (and damned uncomfortable it was too). The fireworks, rock band, a raffle and the after-dinner karaoke, ensured that most of the village came to watch what was happening. Some brought their three-legged wooden stools with them.

Our friends Graham Body, Marina and boys came with us to visit the vineyard in November. This was the first occasion on which we had actually slept at the castle. It was warm – the underfloor heating was working. Hot and cold running water came out of the taps. The bed was pretty comfortable. After a good night's sleep, I was woken by the silence and a strange light creeping round the shutters. Overnight it had snowed heavily and the countryside, which turns brown in the winter, was blanketed in pristine white. The children, who had never seen snow before, enjoyed running around in it, until Daniel threw a snowball at Eliza. The unseasonable snow put an early end to external construction and was the harbinger of further heavy snow falls to follow.

Given the cold weather, and lack of advertising, visitors to the castle after the opening were relatively few, mostly consisting of officials, who "signed" for the meals and wine consumed (we were able collect on these IOUs from the town hall at each month-end). One was so impressed he came back the next day with his mistress.

With such "income" and the sale of whisky to suppliers (a tartan-lined wooden box was designed so that the whisky could be used for the gift season) revenue topped the RMB 350,000 minimum targeted.

CHAPTER 9

THE FIRST BOTTLE

That which costs little is less valued.
Don Quixote

I WAS GOING to use a darker chapter title, but, looking back, I see I have done that rather a lot. So let's accentuate the positive. We bottled our first vintage (2009) on November 11th, 2010. In Beijing on the same day, my fellow Brasenose College graduate Prime Minister David Cameron was asked by his Chinese hosts to remove from his lapel a Remembrance Day poppy, as it reminded them of the Opium Wars. I am pleased to say that he refused. Our first wine to be bottled was a Marselan/Merlot blend from 2009. We had already had a trial run earlier in the year with 1,000 litres of Mark's wine, imported from Australia. But this was the first bottle of our own wine. I shudder to think what it cost. There were only 5,500 bottles in the batch and we did not have a label, so we had to fake up some pictures for a Mail on Sunday journalist who had, bizarrely, found his way from East London to the Scottish Castle in Shandong. He wrote the typical "eccentric Yorkshireman" story.

There is a story behind the first label. During the summer I was visiting Edinburgh's Modern Art Gallery and, in the shop, picked up a book on Scotland's national photographic collection. Flicking through the pages, what should I find but an old sepia shot of a colonial officer, solar toupee and all, toasting a Chinese imperial official. On closer inspection, the colonial officer turned out to be Sir James Stewart Lockhart, Commissioner of Weihaiwei, an old British colony now called Weihai not far from the vineyard, toasting the 76th descendant of Confucius. Perfect! I tracked down the photo to the Lockhart Collection and wrote a pleasant email to the chairman of the trustees, who turned out to be the headmaster of an Edinburgh school, explaining who I was and asking for permission to use the image in return for some cases of wine. This was the reply:

From: "Gareth Edwards" <g.edwards@fc.gwc.org.uk>

To: "Chris Ruffle" <chris_r@heartland.com.cn>

Cc: "Katrina" <katrina@sizedesign.co.uk>

Sent: Wednesday, September 09, 2009 2:40 PM

Subject: Re: Fw: The Stewart Lockhart Collection 4.9

Dear Mr Ruffle

Thank you for your email and explaining in more detail your business venture.

I have now taken soundings from the trustees of the collection and I have been advised that, on reflection, we would prefer not to have the images from the Stewart Lockhart Collection used in this way.

I am sorry to disappoint you, but I do wish you
well with the venture.

Yours sincerely

_____

Gareth Edwards
Principal
George Watson's College

When I enquired about the reasons about the decision (Lockhart was a teetotaler?) I received the following:

From: "Gareth Edwards" <g.edwards@fc.gwc.org.uk>
To: "Chris Ruffle" <chris_r@heartland.com.cn>
Sent: Thursday, September 10, 2009 6:47 PM
Subject: Re: Fw: The Stewart Lockhart
Collection 4.9

Dear Mr Ruffle

I am not at liberty to comment in detail
on the decision though I can state that the
collection has not been used for commercial
ventures in the past. In this, first, instance
we reviewed the pros and cons in relation to
certain other issues regarding the collection,
and it was decided not to pursue this. I
appreciate that this is somewhat vague.

However, I am happy to assure you that it
is in no way a reflection on your particular

venture.

Gareth

Yours sincerely

———————————

Gareth Edwards

Principal

George Watson's College

If you have reached this far in the narrative you will have realized I am nothing if not persistent. Unwilling to be brushed off, I contacted an old friend and one-time colleague James Dawnay, who is well-connected in the Scottish art scene. After receiving a similar rejection from Gareth, James contacted the Lockhart family, who were happy to have the image used. So, after a small donation, I received my final communication:

10 February 2010

Dear Mr Ruffle,

On behalf of the George Watson's Family Foundation, I write to thank you very much indeed for your generous donation of 1,000 pounds, which will be allocated to our funds to further the teaching of Chinese at George Watson's College.

I wish you well with your plans for the Treaty Port Vineyards and I hope that the 1903 photograph will be a successful element in the popularity of the wine.

Thank you also for the kind invitation to visit the vineyards some day. China is somewhere I would like to visit at some stage. Equally, if you are back in Edinburgh and would like to visit us here to see how we teach Chinese and promote our links with China you would be most welcome.

With very best wishes,

Yours sincerely,

Gareth Edwards,
Principal

Victory!

After bottling the Marselan/Merlot, we still had 3,000 litres of Merlot from the upper slopes to bottle. Mark thought this was our best so far and would benefit from a few more months in barrel. I decided to call this "The Admiral", having found an amusing sepia print of the British admiral in charge of the China station, based at Weihai, Sir Edward Seymour, sitting with the famous Chinese statesman and modernizer, Li Hongzhang. We also had 3,800 litres of white wine from 2008 and 2009, which Mark was waiting for some new fruit to freshen up, and 6,300 litres of second-grade red. We also decided to render 11.5 tons of undrinkable wine from pervious vintages into three tons of wine spirit, which was put into the old Marselan barrels. This might come in useful if we wished to produce a fortified Port-style wine in the future.

For the first time the castle, this working castle, took on the distinctive smell of fermenting grapes. It was a pleasure to sit in the library and hear downstairs, behind the ticking of the grandfather clock, shouts and the whirr and clank of machinery; the sound of wine being made.

Before I start on the bad news, I am determined to record all the good. We installed a stand-by generator and larger transformer to stabilise the electricity supply. Both were shielded by a stone wall (design by Ian Begg) to blend into the other outbuildings.

The huge water tank was linked with an additional well to ensure our water supply. A herb garden was begun behind the castle, near the kitchen door (seeds not found in China were sourced on my transits through Amsterdam airport). A formal garden was established along the South wall of the castle, with shaped hedges, wisteria and bamboo. The farmhouse was refurbished to provide more and better accommodation; guests can now sleep on a *kang* (a raised, heated platform used in local farmers' houses).

The castle staff organised several events with growing professionalism: Christmas dinner (with white-bearded Father Xmas himself); A celebration on the night of the Qiu Chuji festival (with moustachioed Taoist priest); a summer barbecue (attended by old college friends Cindy Gallop and Lonnie & Sara Henley); a moon-viewing dinner (with a prize for the table with the best poem about the moon, the surprisingly good results preserved for posterity in the leather comments tome in the library) and a Guy Fawkes' Night bonfire with parkin (ginger cake) and treacle toffee. All accompanied, of course, by Treaty Port wine straight from the barrel. On Guy Fawkes Night, Hao underestimated the power of the fireworks, placing it dangerously close to the building and sending guests diving for cover. At least the castle proved resistant to violent assault.

Our pricing disputes were resolved with Renhe Construction (quote RMB 6.5 million, actual cost RMB 8 million plus lots of hassle) and the decorators, Jitailong (quote RMB 1.5 million, actual cost RMB 2.2 million, but they did a decent job). During the negotiations we discovered that our "friend", Mrs. Ren, now commemorated in the stair mural, was wanting Jitailong to charge us an additional RMB 400,000 so she could use it to pay back our RMB 400,000 loan!

Now for the bad news, which can be simply stated. In 2010, we did not harvest a single grape. The wet summer, which caused poor harvests around the region, was compounded by mismanagement. In the crucial period from July, the ground grass was allowed to get away. Many of the bottom wires had again not been raised to the required height. And the management team of Kang and Hao continued to prefer to react only after disease or insect attack has shown itself, which is usually too late, and preparatory spraying was therefore again neglected. The agreed spraying plan was given no more than lip service. So the vineyard was devastated by mildew. It was such a horrible sight to see acres of leaves browning prematurely and wizened bunches of unripe grapes, rotting on the vine. I began to doubt whether this area was really suitable for making good wine. I also doubted my ability to manage an agricultural enterprise remotely, given the local incompetence and graft. Perhaps I should sell? Some people from Hainan Airlines had visited and vaguely indicated that they might be interested. Tiffany was keen; after her early support she had now come to regard the enterprise as a hopeless money pit. However, in fairness, it should be said that she has the same view of all investment in China. Her grandfather, a landlord in Jiangsu, had his ancestral land confiscated by the

Communists; it was 300 *mu* in area, exactly the same as our vineyard.

Just to top off this disastrous year, I was sent a clipping from the *South China Morning Post*. Jeannie Cho Lee, "first Asian Master of Wine", had tasted some of our wine straight from the barrel and wrote the following snide comments:

> The area is devoting most of its agricultural land to grapevines. Rather than finding inspiration from the modern wineries built along Highway 29 in Napa Valley or grand Bordeaux chateaux, these newly created wineries are clearly looking at amusement parks for inspiration.
>
> I visited one newly built "castle" winery, not because of its eccentric design but because of its location - they are neighbours to the vineyard site currently leased by Domaines Barons de Rothschild (DBR), owners of Chateau Lafite. Treaty Port Vineyards is a recently completed Scottish castle-inspired winery designed by a Scotsman named Ian Begg.
>
> From the tapestry to the dark antique European furniture, the ambience inside the "castle" is very gloomy and medieval. Its owners, a Scottish businessman and his Taiwanese wife, are hoping to attract tourists to this uniquely-themed winery. With six guest rooms and a large banquet hall, domestic visitors can get a flavour of Scotland without having to make the trip.
>
> I tasted their 2009 marselan and merlot blend and found it woody and devoid of fruit. This

light-bodied wine would have benefited from gentle handling and no new oak barrels, but 100 percent new barrels were purchased to mature their inaugural vintage. Much to my surprise, Judy Wang, the Taiwanese vice-chairman of Treaty Port, told me that in 2010 no wine was produced. The winery planted vineyards in 2005 and hired consulting winemaker Mark Davidson, owner of Tamburlaine Wines in the Hunter Valley, Australia.

When I queried why there were no 2010 wines, Wang replied that the summer and autumn brought so much rain and humidity that the grapes were infected with rot and other fungal diseases. This news is not encouraging for its neighbour, DBR, which will be planting its vineyards in spring, less than six months from now.

Well, you cannot please all of the people all of the time...

On one occasion, we were visited by my children, Thomas and Matilda and an old friend who I had known since working together for Warburgs in Tokyo in the 1980s. He was the self-proclaimed godfather of Matilda, despite being Jewish. Whilst on this visit, on a tour around Yantai, he went into anaphylactic shock after eating some shellfish. I was familiar with Yantai hospital, having recently been there when Tiffany had contracted measles. So we managed to get him over to the Yantai hospital emergency room in time and saved him. Having retired from the brokerage business in Asia, he was unemployed. I felt a little sorry for him, so I employed him the following year to help me set up an office for my new fund management business in San Francisco. He duly assembled a group of his friends but, at the crucial moment, tried to blackmail me to improve his already generous terms and, when this failed, pulled out the whole team. He then used Californian labour laws to sue me for unpaid wages. To avoid a time-consuming and expensive court case (expensive for me; his lawyer was working for a share of the winnings, mine was on $1,000/hour) I was forced to pay $400,000 for his betrayal. Truly, no good deed goes unpunished.

登 望 邱 山 綠 樹 蔭
龍 飛 鳳 舞 雲 翳 低
紅 火 興 隆 美 名 揚
酒 香 四 溢 飄 中 西

90

Qiushan valley as it was at first sight

The Mulango Ballet

Ian Begg with local granite

Dairsie Castle before

Dairsie Castle after

Tiffany with the builders

Planting 200 mu

Builders Wang & Sung with Eliza, in front of the still unfinished castle

The team in early 2008

Eliza samples the cherry harvest

The team in late 2008

Tending the vines

Mark Frazer and Chris find something to laugh about

The Great Hall

Roman & Ewe and the main stair mural

With Matilda, Thomas, staff and old friend

The cheerful team at our first bottling

Hao misjudges the fireworks

Photo of Chris (the one in the kilt) toasting Penglai Mayor Zhang

A plan of how our 300 mu (21 hectare) vineyard looked by mid-2007

Eliza, Judy and rubber cheque

William, Chef Rainbow and claret Aga

Some of the crowd at Philosopher Qiu's birthday

Opera singers off duty in Mulangou

Appropriately dressed for a castle opening

The best of the Merlot being hand sorted

Edward Davidson helps, briefly, with sorting

A difficult lunch, with accountant Sun, John, Judy, Tiffany, Chris and Emma

Winery Manager Hans Zheng, with the first in-barrel sales

Grenache in flower, before rain request to Taoists

Where the wine is made

Pests in the Grenache…

The stars revolve around Qiushan

The castle under snow

The London taxi awaits late diners

The Scottish Castle from the top of Qiushan

CHAPTER 10

FROST ON SNOW

Foied Vinom Pipafo Cra Carefo
(Enjoy the wine today; Tomorrow there'll be none)
Etruscan inscription

IF THIS WERE a work of fiction I would now tell the reader an uplifting story, to vary the pace. But instead we just had more trouble piled on trouble or, as the Chinese saying goes, "frost on top of snow". Literally, in this case. After the terrible summer of 2010, the winter was the hardest in sixty years, with extended periods of sub-zero temperatures. Many already-weakened vines were killed by the cold. By the first rain in May we had lost 21,000 vines, with Grenache, Syrah and Cabernet Sauvignon being the worst affected. Much of the rest died back to the roots, meaning that there would be no fruit from 70 percent of the vineyard again this year, just the extra work of retraining the vines. In the spring, manager Hao, in his normal gloom-laden tones, predicted that the crop would be no more than 5 tons. Mark dismissed this, saying that, with good work, we would harvest 25 tons. In this case, one of the rare ones, Hao was closer to the truth. In the end we harvested just 7.9 tons (we should be producing 60 to 80 tons).

Grape Production 2011

Date	Variety	Pick weight 收获重量	Sort weight 筛选后重量
9.08	Grenache Noir 黑歌海娜	268	229.1
	Cot 高特	300	259.5
	Sangiovese 桑娇维塞	80.6	73.45
	Muscat 小白玫瑰	34.1	6.95
	Cabernet Franc 品丽珠	59.25	65.15
	Arinarnoa 阿娜	139.3	123.8
	Merlot 梅乐	733.9	711.04
	Chardonnay 霞多丽	27.6	12.45
9.09	Merlot 梅乐		3218
	Marselan 马赛兰		973.7
	Cabernet Sauvignion 赤霞珠		719.4
	Petit Verdot 小味儿多		1,476.5
Totals/Kg 合计		1,642.75	7,857.04

Despite this miniscule production, costs had sky-rocketed. Total costs in 2011 reached RMB 3.3 million (excluding the substantial fees paid to Judy, John and Mark) up from RMB 2.4 million in 2010. This partly reflected the extra work of raising the bottom wires (still not fully completed) and replacing rusting wire, as well as wage inflation (running at about 15 percent in China). But it also reflects weak management of contracted labour. Kang left for another job in March. Hao was given a chance to manage the vineyard by himself, but proved an office farmer, and stubbornly resistant to Mark's advice, so we fired him after the harvest. The employee turnover at our small enterprise remains very high. Oh, for someone reliable and competent.

My day job (China fund management) was also very difficult that year. In addition to poor markets, the company I worked with for 16 years, chose to vilify me in a legal dispute, ending our co-operation in May. So with my partner Ke Shifeng, we decided to buy out our erstwhile employer from our joint venture and set up our own company. Consequently, I spent a lot more time (and money) with lawyers during the year than I would otherwise have chosen. So I had little time to spend on developing the wine business. Sometimes the two businesses seemed to blend. The actual blurb on the back of a bottle of Castle Red reads:

Left to the tender mercies of a compliance officer in the financial industry the label would probably read:

> "This wine may, or may not, be red. Risk warning: this product includes alcohol. Imbibing this product may lead the purchaser to become intoxicated. The company can accept no responsibility for any actions taken after consumption. Consumers unsure of their rights with regard to the drinking of wine should take legal advice."

We launched our wines at the Top Wine exhibition in Beijing in May, and also attended exhibitions in Shanghai and Yantai later in the year. Our booth was small and in a distant corner of the main floor. But it looked smart, and the strategy of male staff wearing kilts certainly attracted attention. At the Yantai exhibition, where we were helped by a be-kilted Matthew Kimber, the handsome son of an old friend, we could have sold more autographed pictures of Matthew than wine.

The comments we received from visitors were generally good. "Is this really Chinese wine?" was the most common. We sold a few bottles here and there, so became very excited when John and Judy reported that they had been contacted by a company buying for the People's Liberation Army who could, for patriotic reasons, only drink Chinese wine. After some negotiations, and one extremely expensive dinner (RMB 4,000), the triumphant couple returned with a signed contract worth RMB 1.1 million, buying a large shipment of our reserve wine at full price.

Of course, it all turned out to be a confidence trick, taking advantage of bright-eyed and naïve newcomers. The other party did all this just to split the cost of the dinner and some expensive bottles of *baijiu* with the restaurant owner! At least we were wise to the scam when a group from Ningbo attempted something similar. Overall, our experience of taking stands at wine fairs was not encouraging; we just seemed to be paying for a lot of people to come to drink our wine for free. Few real leads resulted.

In the end, our total sales for the whole year, including accommodation, only amounted to less than RMB 1.3 million, or under 5,000 bottles. It is a good thing that we are selling wine and not bananas and that the inventories in warehouse can keep. In addition to the 2009 vintage, we also have stores of a 2008 Banyuls, a natural sweet wine from the Southwest of France, which used to be popular in Victorian England and is the only wine I find you can still taste after eating *mapo doufu* (spicy beancurd). There is also a 1998 Appollonia, a sweet wine from Cyprus with an interesting nutty taste that I have found goes well with Taiwanese dishes such as Ludan. It is the closest grape equivalent to Shaoxing wine, yellow wine made from fermented rice and popular in Eastern China. Both of these were imported in bulk and bottled in the winery. Because of their relatively high alcohol content, both of these wines travel well (an important consideration as Chinese transportation and storage can be a little rough).

There is also a large store of Whisky Mc. This is a traditional cocktail invented, as legend has it, by a corporal in the Indian army called MacDonald. It was a personal favourite of mine when entering a pub from inclement north British weather conditions. In our iteration it consists of imported 3-year old whisky and Crabbie's ginger wine from Scotland, blended and bottled in-house. I find that it appeals to the Chinese taste, sweetens up the whisky with the recognizable taste of ginger, so you can still *ganbei* it (knock it back in one)!

Setting the price of our wine was an interesting challenge. It is one of the relatively few industries where the price of the product is not necessarily linked to the cost of production. The "story" or brand is all important. China is a market which is used to paying high prices for wines, given the history of high import duties (now "only" 40 percent in total). And, given the lack of wine education, a high price wine tends to be seen as "better" than a cheap one. Given our low volume and high overheads, we need to sell as a premium product. We therefore set the retail price of our two 2009 reserve wines (the Commissioner and the Admiral) at RMB 500 and the price of Castle Red, Castle White and Whisky Mc at RMB 180, with appropriate discounts for distributors, volume purchases or castle guests. The key will be to sell as much as possible directly to consumers, rather than middle-men, at the retail price.

Of course, there was no 2010 vintage to be launched. In barrels, we just had 900 litres of a rose, to be released in the spring of 2012, and 3,900 litres of a blended red (Petit Verdot/Marselan/Merlot), but that would not be released until after barrel ageing in 2013. There was also 5,000 litres of an experimental port, which Mark was concocting using our own brandy.

Despite the losses, I have been an effective pioneer in terms of attracting other investors. Lafite went ahead and planted its neighbouring plot, after a one year delay.

Eventually they had to source the bulk of the vines locally, rather than importing from France, due to stiffening quarantine restrictions. The Taiwanese private equity investor K.C. Kong also planted around the quarry at the South end of the lake.

Meanwhile, the boss of Daxindian's largest company, the cable maker Pengtai, decided to get in on the act. He cleared a huge area down by the lake, and behind Mulangou, for a vineyard, showing Lafite's "exclusion zone" to be a worthless piece of paper. I feared a hideous concrete Chateau, which he will surely want to build next to us. When I visited the boss in his office, he kicked just such a plan behind a potted plant in an attempt to avoid my seeing it. There seems to be no concept of planning control or environmental protection yet in China. This year the hills along the skyline have been covered in windmills; the local government has "new energy" targets to meet. At a big political lunch for 45 people which I hosted in December (you have to keep working those *guanxi*) the cadres were delighted to tell me that in addition to the new airport which would be built just fifteen minutes away from the castle (opened in 2014), a connecting motorway is going to run across the valley, right in front of the lake onto which we look. Wouldn't it be convenient? The mayor's speech explained to me how much I owed the Communist Party. I am promised deeds to the castle, municipal rubbish collection and road signs "soon".

After the harvest we decided to reduce the size of our vineyard, tearing up those plots which are too disease-prone to be worth cultivating. These tend to be in the valley bottom, or behind apple orchards, shielding them from the drying wind and increasing humidity. We also abandoned a nice sloping field on the East side of the hill, because it was just too remote, and tended to get "forgotten" in spraying . This will reduce the growing area by 36 *mu* to 278 *mu*. (We planted sweet chestnut and walnut trees in the vacated land, on the basis that they would not require too much looking after. This did not work as they were not looked after, and a fair number were stolen). Once surviving vines from the abandoned areas have been re-planted in ongoing plots, we still need over 18,000 vines to make good the ravages of weather and mishandling. Given the expense (now RMB 8 per shoot) I decided to build our own nursery, so we could use (or sell) our own cuttings, and a greenhouse was erected in the farmhouse yard. Other constructions completed this year were a Scottish sundial (slightly crooked), a seat in the herb garden, and a little park with benches directly in front of the castle.

We had a growing number of visitors in 2011. My social standing rose in October when I arrived in Yantai in a private jet. This was owned by fund investor Don Campbell and was my first such trip. I felt like a rock star strolling through the VVIP entrance at Shanghai Pudong airport. Unfortunately the minibus journey out to the plane, parked in a distant corner of the vast airport, took 30 minutes, and the plane was routed right around the Shandong peninsula (for strategic reasons?). So a commercial flight would have been quicker (but less fun). Don gave a lift to my old friend David Erdal, a fierce proponent of worker ownership, and his authoress wife, Jenny, which led to some interesting dinner conversation. She later reported her trip in the FT.

My college flatmate, William McGrath, came to visit the only AGA oven to be installed in China so far (he is the CEO of the company). AGA also supplied the matching fridge (the chosen colour being "claret", what else?) as well as three wood burning stoves; burning apple tree prunings keep us warm in winter and make the castle smell as it should. We also imported some nice hand-made tiles from AGA's Fired Earth subsidiary.

A variety of journalists arrived to drink our wine, and some of them published stories about us (including articles in the *China Daily*, and the ANA in-flight magazine.) In the interests of further publicity, I pretended to be a wine expert at a meeting of the Foreign Correspondents' Club in Beijing, as arranged by our friendly Irish journalist Mark Godfrey. We even went into publishing ourselves with two editions of the Treaty Port Times:

Jennie Erdal

21/22 January 2012

In this Chinese year of the Black Water Dragon, a creature imbued with the utmost symbolism, Chinese people are encouraged to be optimistic and to take risks, to get married and to start businesses. The energy and vitality of the dragon are said to favour boldness, not humility. On my recent visit to China, there were posters warning people against "fake dragons", exhorting them to buy only genuine dragons, not dinosaurs or other inferior creatures. The dragon, according to the slogan, is "wise and wealthy" and, above all, "smart" but also unpredictable, evidently on account of its head and tail not being visible at the same time. Perhaps not unlike the unbridled beast that is the Chinese economy.

Being able to tell the genuine from the fake must be one of the greatest challenges for upwardly aspiring Chinese, surfing the unstoppable rise of capitalism. *Shanzhai* – imitation goods – are everywhere. In Shanghai, the counterfeit capital of China, I was invited by a visiting Californian on a spree to a huge covered market that specialises in fakes. It was meant as a kindness, and I had wanted to show willing by getting into the shopping-till-dropping groove. But faced with free enterprise on such a hedonistic scale, I felt queasy and paralysed, unable to buy anything. The Californian had no such hang-ups. Through our interpreter she bargained and battled – "I'm keeping the Chinese economy afloat" – ending up with irrational quantities of fake high-end swag.

To arrive in a country without knowing a word of the language is a strange and unsettling business. Instead of delighting in the local conversation, and joining in when you can, you are unhitched from your surroundings, moronically marooned. It is also a useful lesson: it forces you to look into people's eyes, interpret their gestures, the intent behind the bewildering sound stream. I was lucky enough to travel to China with my husband, a Chinese speaker who taught in Tianjin at the tail end of the Cultural Revolution. For the most part we were separated only for short times but in Beijing I had to spend a whole day alone, wandering wordlessly around the *hutongs*, the narrow backstreets seething with life: pairs of women sitting on doorsteps and playing the ancient board game of Wei Qi, others gathering round to watch, while those on the move all carried something – bundles of laundry, stacks of cardboard, other people's waste, even a coffin – on a staggering variety of bikes and handcarts. Kipling thought that the first condition of understanding a foreign country was to smell it and in those clotted alleyways, with the pall of pollution, street food and dodgy drains, there was ample scope for understanding without the need of words.

In Shanghai we stayed with an unusual couple – a brilliant investment fund manager from Yorkshire and his Taiwanese wife. They had arranged for us to visit, as part of my husband's research project, a handful of companies where profits are shared with the employees. Worker ownership is enlightened thinking but it's hard to imagine it catching on in corporate China. Despite the key part they play in the country's economy, most migrant workers are denied even basic rights – largely on account of the *hukou*, an ancient residency status system that is often referred to as "China's apartheid". Though in direct ideological conflict with communism, it was strongly enforced under Mao, and even today, in spite of growing unrest, there is little prospect of change – unless by some miracle the *hukou* itself were to become an impediment to economic progress. At which point, in a country that already has fake Apple and Ikea outlets, the unpredictable dragon might foster fake John Lewis stores, but with employees properly rewarded and valued in a genuine partnership culture.

At a clothing manufacturing company, whose brand concept is designed to appeal to sharp young professionals, the presentation – in English – was entitled Passion for Fashion (pronounced "Peshun for Feshun"). Afterwards, during a Q&A, when asked why all the models in the visuals were European, the chief

executive said: "There is a simple rule: if you want to keep the price high, don't portray your brand as Chinese."

Our friends, as unpredictable as the dragon, have been busy flouting this advice. Some years ago, while entertaining an American client from the Napa Valley, they asked if he would like a Chinese wine to accompany his meal. "Ask me again in a hundred years," came the wine-snobby reply. Thus was the seed of an idea planted, now flourishing as 85,000 vines imported from Bordeaux to a beautiful valley in the eastern Shandong peninsula - the bit that sticks out into the Yellow Sea, pointing towards Korea and Japan. The marine climate is favourable, the land fertile and, significantly, it is on the same latitude as the Napa Valley.

Since every self-respecting, wine-producing estate needs its own château, plans were duly drawn up. Not, in this case, for a baroque French pile but, surreally, for a traditional Scottish castle, built from granite and complete with turrets and battlements. After protracted meetings with party secretaries and assorted officials, much banqueting and toasting plus a good measure of *guanxi* - the essential "connections" for doing business in China - contracts were signed, speeches made, fireworks lit and *feng shui* carried out on site. After which two local builders, Mr Wang and Mr Sung, were flown to Scotland for a crash course in castles. Our friends became major employers in the region. In the nearby treaty port of Yantai, the official history in the old colonial quarter cites the exploitation of the Chinese by foreign imperialists. But in the castle and surrounding vineyards a new orthodoxy prevails: foreign capitalists creating local employment through investment.

Our stay at the Scottish castle in Shandong was like turning up in someone else's dream: lots of beautiful dots that were hard to join. A flight in a private jet called Chérie, meet-and-greet on the runway by kilted Chinese in a London cab, more tartan and tweed at the portcullis, whisky in the library, suckling pig in the Great Hall. All smoothly blended with fine wine - reserve Merlot, from grapes hand-picked on the upper slopes of the estate and matured in French hogsheads.

On our last day we climbed a nearby hill and looked out over the valley. The magnificent castle, flying the saltire, nestled beneath, while workers in old Mao suits tended the vines. Inspired undertaking, we wondered, or the folly of a mad Yorkshireman? The answer lay a little to the north on adjacent land, where more terraces are under construction. The new neighbour is Château Lafite.

Jennie Erdal's novel, 'The Missing Shade of Blue' (Little, Brown) is published in March.

I gave a nerve-wracking talk entitled "How I came to build a Scottish Castle in China" at the TED Global conference in Edinburgh in July, accompanied by the first-ever tasting of Treaty Port wine outside China. I had shipped a pallet load to Yorkshire at great expense, which allowed me to supply this tasting as well as various family, friends and helpers (including a case for the great-grandson of Sir James Stewart Lockhart). I have a grainy, mobile phone photo which shows me opening the first ever bottle of Treaty Port wine offshore. This took place in Hong Kong in the smart Italian restaurant Domani, owned by my friend J.R. Robertson, and assisted by the CFO of luxury product distributor Sparkle Roll (yes, Chinese-listed companies come up with some cracking names) and CITIC broker May Lee.

The number of pilgrims celebrating Qiu Chuji's birthday in deep mid-winter had grown every year. From a small bonfire and few fireworks around midnight when I first arrived in Mulangou, thousands this year visited throughout the day, and were entertained by an enterprising

TREATY PORT TIMES

Quality Wine from China

NOVEMBER 2011   VOL. 1 ISSUE 2

Harvest Time 2011:

An unusually harsh winter made for a difficult 2011 harvest at Treaty Port. "Volume of only 9 tonnes was disappointing," explains Treaty Port CEO Chris Ruffle. "The season itself was reasonable, but the damage had already been done by the harsh winter which was the coldest in 60 years. This killed some vines and caused many more to die back to their roots. These are now recovering, but produced no grapes this season.

The best crop came once again from the Marselan and Merlot varieties, though we had a decent crop from Petit Verdot for the first time. Our intention is to produce from this seasons grapes a Marselan/Merlot blend

(probably for release in 2013) and a Rose.

We continue to learn about the climate, our vineyard and the suitability of grape types."

Treaty Port staff in their stylish Scottish kilts

Treaty Port, said Ruffle, would continue to experiment with higher trellising, ground cover and spraying methods to find the best combination. "Unfortunately each experiment takes a year and is subject to the weather. Fingers crossed for a better winter…"

Treaty Port Vineyards Limited 登龙红酒

Mulangou, Daxindian, Penglai, Shandong 265612, China

www.treatyport.com

Shandong opera troupe. Rumours are that the Qiu Chuji temple will be rebuilt with funds from local businessmen. "I am a believer," said the Party Secretary in a quiet aside. "But of course the Communist Party cannot play any official role." The interest by the Chinese people in recovering their history, long buried by Communist party propaganda, is clear.

My sons, Thomas and Walter, came to stay for Easter and we sampled the Aishan natural hot springs for a spot of R&R. This is the strange, three-fingered mountain which can be seen on clear days from the castle. It is about 40 minutes drive away. It is worth a visit, but is a spa with Chinese characteristics. There are a series of hot pools flavoured with various vegetables and Chinese medicines, which are good, apparently, for different parts of the anatomy. We were driven by "Big Che", who bears the best resemblance to a real London taxi driver I could find. He claims to be descended from Genghis Khan, but then again, who isn't?

I finished writing this chapter on Christmas Eve 2011, and was not sad to bid farewell to a tough year. One unlooked-for benefit is that my difficulties brought me closer to my family and proved who my real friends are. Perhaps 2012 would be better? Anyone reading this far already knows that I am, as any foreigner staying for long in China must be, a dyed-in-the-wool optimist. I was asked by an American journalist to name ten qualities required to set up a vineyard in China. My answer for the record is:

- Vision (Why would you do such a thing? Are you crazy?)
- Deep pockets (no cash flow for six years)
- A good liver (for drinking *baijiu* with party officials)
- A strong stomach (for eating sea slugs, donkey and camel's paw with aforementioned officials)

- Trusted local advisors (to explain a way though byzantine bureaucracy)
- A source of good quality vines (imports now restricted)
- A stubborn streak (for when things go wrong – as they will)
- A sense of humour (ditto)
- An ability to entertain oneself (the Chinese countryside at night is very quiet)
- An understanding wife

And finally, I was asked to write a poem with just six words about the ingredients needed for building a Chinese vineyard:

Grapes, patience, persistence.
May contain sulphites.

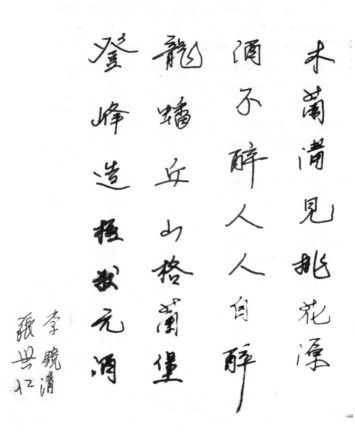

源 花 見 溝 蘭 木
醉 自 人 不 酒
堡 蘭 人 醉 龍
酒 格 山 蟠 登
　 狀 極 丘 峰
　 元 造

INTERLUDE I
A LETTER FROM CHINA -
THE CHANGING FACE OF CHINA'S
COUNTRYSIDE
2013

MOST PEOPLE WILL BE aware of the huge changes that have taken place in China's cities over the past twenty years; even James Bond has used the futuristic towers of Shanghai as a backdrop for his brand of mayhem (though in the Chinese version he doesn't shoot the Chinese guard). Less well known, perhaps, are the changes taking place in China's countryside. This is not surprising: foreigners' visits tend to be confined to the big cities, and even local commentators can be rather urban-centric. However, it is important given the scale involved, a scale that is evident from the window of any Chinese high-speed train, as hour after hour of well-populated rural landscape flashes by.

But I am not going to write about macro trends. I am going to write about one small village called Mulangou. The name can be translated as "Peony Gully", but also summons up the ghost of the famous woman warrior hero Mulan. I first visited Mulangou in 2004. As a personal investment, I rebuilt a farm there in 2005 and have been back regularly ever since. Even using Shandong's fine and lightly trafficked motorway system, Mulangou is a full hour's drive from Yantai, a minor port and the local administrative centre. When I first saw Mulangou it had only a mud road. After a heavy rain the tricky slope up from the neighbouring village became impassable to vehicles. Neither were there any street lamps; when the sun went down it was very dark – nice for star gazing, but little else. Nearly everyone was (and still is) called either Huang or Gong; I made the classic error of befriending the village head Huang and thus immediately alienating the other half of the village. The population seemed to be made up of the very old (or prematurely aged by working in the fields) and the very young. Men's clothes were the standard blue cotton or khaki, and transport was the walking tractor, ancient bike or muddy motorbike. A car meant that a party official was visiting.

Progress was initially gradual – a distant reflection of the Hu/Wen administration's efforts to improve the lot of the farmers by boosting food prices and improving social security. When village head Huang suffered his nasty motorbike accident the burden of treatment fell almost entirely on his family. Today Mr. Huang would have received some help from the national rural health-insurance scheme, but that was not in place in 2005. The main road was paved by 2006 (and lined with trees). I contributed some funds to buy a few street lights and was then sent the bill for the electricity. The makes of motorbikes were upgraded. Some farmers bought proper tractors. The quality and variety of younger people's clothing improved. But not all progress is positive. Litter became a terrible problem. Previously, farmers' garbage was largely organic, but now they were shopping at supermarkets. With no rubbish collection, discarded plastic bags and packaging blew across the fields and snagged on the apple trees.

We cut forward to 2013 and to some photographs taken in Mulangou last week. The photo below may look dull at first glance but look again. Solar-powered street lighting! (China has a surplus of solar cells to make use of). And a garbage bin – the local township has now started collecting the rubbish. You will also note that a number of side roads have now been paved.

The number of cars in the village has increased many-fold, especially on the weekends. A car is no longer a sign of a visiting official (though this is still a good bet if it is a black Audi) but of owner-drivers from Yantai or Penglai who are making a family outing to the countryside. A number of the more enterprising farmers now offer genuine countryside lunches, where the visitors can eat very well for 50RMB. A little garden has been laid out in the village centre with, bizarrely, a full set of gymnastic equipment with instructions in English. None of it is ever in use when I pass.

The village is now a veritable hive of construction activity. A new government-sponsored irrigation system has been installed. The hills are dotted with windmills (it is unclear whether any are connected to the grid). A new village hall is being constructed to replace the old one, on which faded red characters exhorting the farmers only to have one child can still be made out. This is an object of pride for the new village head (the only political position in China for which candidates are chosen by popular election). More esoterically, the local businessmen have sponsored the rebuilding of a Taoist temple, knocked down at the time of Liberation. The philosopher so honoured, one Qiu Chuji, who lived here in the thirteenth century and advised Genghis Khan, has his birthday in January. When I first arrived this was celebrated by a hundred or so villagers, a bonfire and some fireworks. Now it is a two-day festival, with opera performances and Mulangou's only traffic jam. The local party officials regard the temple as nothing more than a tourist attraction, but I think something more interesting is happening here. I cannot claim that there have been any major advances in agricultural technology or mechanization – the close-planted apple and peach trees do not lend themselves to this. But there seems to be a revival of pride. Although the process of urbanization will continue, and perhaps even accelerate if the new administration brings in *hukou* reform, the countryside may no longer be just a place from which to escape.

CHAPTER 11

YEAR OF PERSISTENCE

Our own worth is measured by what we devote our energy to.
Marcus Aurelius

IT IS A FINE sight from the top of the castle this morning as I continue this irregular journal. The vineyard is full of snow, lit by a low-angled sun. The frozen lake is so bright it is hard to look at. Last night was a Christmas party attended by a small but enthusiastic group who battled the icy roads to attend. Roast turkey was accompanied by the last of our Castle Red 2009. Our Treaty Port Port made its debut with dessert, party games and silly hats. I notice, reading back, that this journal has a rather wintry theme, which is misleading. It is just that the year-end is when I find time to sit and review the year.

This year, like all the ones detailed above, was a difficult one. My investment business, which funds the vineyard, continued to be under heavy pressure from a weak global economy and investors' fears about the sustainability of Chinese growth. The challenge is made harder by the hostility of my former employer. All this reduced the amount of time I could spend worrying about the vineyard and emphasised the need to cut costs. In

February, therefore, we replaced John & Judy as managers. They had done well during the opening of the castle and had style, but they were expensive, both in terms of salary and all the tickets back-and-forth to Taipei. They were not always present at the castle and cost control was weak. The matter came to a head when Judy submitted a budget showing costs due to climb by 60 percent with little improvement in sales.

We flew up to Shandong in bitter weather to install the new manager, Emma Chen, the sister of one of Tiffany's friends in Shanghai. Although also from Taiwan, Emma had worked in a factory in Dongguan in South China for 5 years, so is aware how business operates here. It was a tense visit and I remember an uncomfortable lunch, cross-legged on top of the kang bed in the house of our lead farmer in the vineyard, Huang Yakuei. Food piled up, until it was difficult to find space on the small table, as conversation faltered.

The visit also happened to coincide with the annual Qiu Chuji celebrations, and, walking back through the village, I met the Shandong opera singers on their way home. They had not yet taken off their stage make-up, but had changed back into civilian clothes, creating a surreal effect.

Emma subsequently did a good job on cost control, cutting total expenditure from RMB 3.8 million in 2011 to RMB 2.5 million in this year. Revenue also rose from RMB 0.9 million to RMB 2.0 million, with wine sales gradually picking up, though we remained very dependent on sales to the local government. I hasten to add that this cost calculation is made on a cash basis, with no consideration of my time, depreciation of equipment or a number of offshore costs (notably Mark's consulting fees). Some necessary repairs and improvements to the castle and vineyard were put to one side and no further wine or whisky was imported, so the figures flatter. Tiffany continued to urge me to sell the whole business, but I could not see who would buy it, especially as we still did not yet have the deeds for the castle. Yes, three year after we completed the castle our application continued to wind tediously and expensively through the many layers of Chinese bureaucracy. The fire department in the distant provincial capital of Jinan remained a road block (apparently they were too busy to come out all this way to inspect our fire safety preparations).

The weather this year was pretty good. Part of the vineyard was still recovering from the ravages of the winter of 2010/2011, but hopes were high for a good harvest. This was an email from Mark written on August 10th:

> The vineyard looks very good - as good as the photos showed. There is a little berry disease (anthracnose) because of humidity in the whites and some syrah. The rest of the fruit looks very good at this stage and better than I have seen at this time in previous seasons…
>
> We are very close to harvesting the white grapes and Emma, Huang and Shao have been very co-operative and we are organised. The analyses of the whites is quite similar, the flavours are almost there, the weather remains humid without much rain in the forecast. Fingers crossed - it has been a drier than normal August so far. We have done an estimate of the possible tonnage of over 50T but given the history here, I stress this is tentative. Remember many vines were

re-trained from the ground after last years'
frost.

The harvest started with the white grapes on August 15th
and was already completed with the last of the red grapes by
September 7th. We actually harvested only 19.8 tons.

Picking Date	种类 Type	名称	Variety	总重量/千克 Weight/Kg	混合后糖度 Sugar content after mixing	调整后糖度 Sugar content after adjustment
8.15	白葡萄 品种 Whites	小白玫瑰	Muscat	2,069.65	15brix	19brix
8.15		霞多丽	Chardonnay	484.75		
8.15		维欧妮	Viognier	508.35		
8.27	红葡萄 品种 Reds	梅鹿辄	Melot	4,500	22.5brix	无调整 no adjustment
8.27		西拉	Syrah	2,041	17.5brix	无调整 no adjustment
8.27		黑歌海娜	Grenache noir	1,176		
8.31		桑娇维塞	Sangiovese	1,175	17.5brix	19brix
8.31		马赛兰	Marselan	2,411		
8.31		小味儿多	Petit verdot	2,038		
9.02		小味儿多	Petit verdot			
9.02		安娜	Arinarnoa	336		
9.07	桃红酒 Rosés	赤霞珠	Cabernet Sauvignon	1,753.9	18.9brix	无调整 no adjustment
9.07		品丽珠	Cabernet Franc	1,313.6		
		高特	Cot	0		
Total				19,807.25		

This is another email from Mark on 9th September:

CHRIS RUFFLE

The tonnage is disappointing Chris. Factoring
in the vines that are recovering from frost
and had little or no fruit, and the number of
re-plantings required, and the areas already
removed, I would have expected around 60 -70
percent reduction in estimated optimum yield
of 110-120 tons (approximately 400kg per
mu). The anthracnose/wilt issue definitely
took most of the difference compared to our
initial estimates, but there did appear to be
some kind of spray anomaly with the Marselan
which is impossible to get to the bottom of and
dramatically reduced that yield to my surprise.

In dry seasons like this, some supplementary
water availability in spring and early summer
would have assisted bunch weights as well - a
luxury we don't have at present.

Now that the long path at TP has seen some
fundamental necessary changes made - better
staff management and compliance, weed control,
proactive spraying versus reliance on a
'curative' regime, trellis/vine height - I will
be focusing on techniques to better control
anthracnose infection.

As we discussed at some length, this is the
first year I feel I have learned much at all
about the applied management effect over an
entire season. That said, the consistency and

thoroughness of the spray regime under Huang's leadership can still not be entirely trusted. Emma worked well in her first year involved in this sort of technical farming and will be much better qualified to lead and supervise next year. She now sees the importance of being equipped properly to do the job and accepts that on a number of occasions this year the spraying was drawn out or delayed.

Syrah and Cot in particular still remain suspect varieties, but I think warrant a bit more patience. In a decent year, both could make a pretty good Rose style, once we better deal with anthracnose. Both are growing back well after the frost issues.

Sangiovese is also suspect.

All this said, the white really shows some good fruit characters and the Merlot post ferment was the cleanest red I have seen at TP ever and achieved adequate natural ripeness of 12.5Be.

I mentioned to Mark that just as some vineyards produce a sweet wine from grapes dried by the Botrytis fungus, or "noble rot", perhaps Treaty Port's speciality could be sweet wine from grapes affected by anthracnose; we could call it Treaty Port Anthrax.

He wasn't amused.

The wines that Mark, assisted by Pip Martin, made from 2012's crop were a fruity white, caused by the preponderance of

Muscat, which we called "A Lady of Fashion" and a semi-sweet Rose, called "The Debutante". These were to be bottled in clear, fashionable half-bottles, but the cap supplier provided us with caps that did not work on our machines, so we had to revert to 750ml bottles. Given our small size, we struggle to get suppliers to pay us any attention, and must make do with whatever standard materials they have in stock. "Treaty Port Port", made with our own brandy distilled in 2009, has been put into a green half-bottle, which takes our standard cap.

I came under pressure to bottle Castle Red 2011 as our whole stock of Castle Red 2009 was bought out by an agent from Henan.

For the first time that year, we made wine for an outside vineyard on a toll basis. The customer was Luye Pharmaceutical, owned by Liu Dianbo, who has been one of the best customers of our wine. He also hired Mark to consult on his own Laishan vineyard project which, truth be told, has a lot more to do with real estate than wine. His ambition is to sell 100 "holiday chateaux" with vineyards attached. I advised him that, given the difficulty of running a vineyard in China (see all of the above) he might want to plan on leaving the country once he has sold out. My impression was heightened when I actually got round to visiting the project – now called Pula Valley; the vines have been planted on small, steep terraces, mainly North facing, with no thought to mechanization or even access. This time we processed 8.5t of Italian Riesling and Cabinet Sauvignon grapes for them. Unfortunately, in concluding the contract, we made the rookie error of basing our RMB 5,000 per ton charge on the volume of wine produced rather than weight of grapes received (one tonne of grapes makes about 600kg of wine), which meant we lost money on the project. My lack of confidence in our managerial skills led me to turn down an approach by Gold Cider, who wanted us to produce cider under OEM contract for them. Whilst it might cover some overheads, I decided it would be too much of a distraction given the continued problems in our main business.

The nursery to develop vines to replace our own losses and potentially sell to others, for which we built a glasshouse, also turned out to be a failure. In the heat of the summer, someone forgot to keep watering the shoots and they all dried out and died.

Now we just bury stronger cuttings in a hole in the ground and plant them out in beds when the weather warms up. The greenhouse stands empty.

The local Pengtai Cable boss continued to be a nuisance. Throughout the summer his lorries continuously carried stone past the castle for the creation of his vineyard down by the lake. He also pushed on with his plan to build a hideous 60-room motel on the hill along from the castle and above our vineyard (no planning permission required if you have local *guanxi* and new money to commit). He widened roads through the vineyard to make it easier for his trucks, actually cutting a corner off one of our plots without so much as a "by your leave". On the other side of the castle, he has been the lead in rebuilding Qiu Chuji's temple. I had originally imagined it would be rebuilt as the small temple which had been there before it was knocked down at the time of communist victory. But local bosses don't think like that – big is definitely better. Whether it will turn into a living temple, or just an empty shell remains to be seen. The original stone stele at which people prayed was, bizarrely, left outside the temple enclosure.

Party Secretary Zhang took me aside to ask if he could take the corner off one plot to make a nice straight road up to the temple. I told them Daoists prefer winding ones. I did not hear any more about it - for a while.

Lafite finally planted their vineyard this year – all red grapes, Cabernet Sauvignon, Merlot, Syrah and Marselan. There hasn't been much interaction with us, but the international technical director, Eric Kohler, did stay at the castle on one visit. He was politely disdainful of our efforts. Lafite's presence is starting to attract other visitors.

This is an email that arrived after a visit by Albiera Antinori:

Dear Chris, Emma and Mark,

Just a quick note to thank you all for the hospitality you offered us in your magnificent Castle in Penglai.

It was really interesting for all of us learning about this production reality and it was really enjoyable spending some time together talking about this incredible country.

Please accept our congratulations on your wine's quality.

We really hope to have the chance to reciprocate the hospitality soon in Tuscany.

Dear Chris,

Your property and project is really amazing and well presented by your staff.

I hope to see you in Shanghai to drink some

```
wines together.
```

```
Best regards,
Jacopo Pandolfini
Asia - Pacific Export Manager
```

Tzyy Wang took over from her husband Mark Godfrey to provide occasional help on PR. She published two editions of the Treaty Port Times this year and organised some local magazine coverage. We attended another wine expo in June in Beijing, near where the Olympics took place in 2008, which is now a weed-studded wasteland. I increasingly doubt the utility of such events, whose primary function seems to be to provide free alcohol to the local populace. Advertising in general is extremely expensive in China, and for a small vineyard does not make any sense, when worked out on a cost per bottle basis. For a few months we ran, at great cost, an advert in the arrivals hall at Yantai airport, but its sole function seemed to be to give me a warm feeling as I was collecting my baggage. Only one guest ever told me that they had seen it. Our other attempt, undertaken at Emma's insistence, advertising on the side of a bus probably did not do too much for our premium brand image. What we need is free publicity from journalists who quite fancy the idea of visiting a castle to drink wine. Perhaps I'll even write a book.

In April, I visited Yinchuan in Ningxia. I took the opportunity to look up the Gao Yuanyuan Winery, which produces a wine I had heard praised called Silver Heights. The plain, sheltered from the Gobi by the Helan Mountains, is 1,200 metres above sea level. The summers are great for grape growing – lots of sun, little rain, no insects. Unfortunately, the winter temperatures get down to minus 20 degrees Celsius, so they have to bury the vines in November, digging them back up again in the spring. Visiting in April I watched as acres of vines were exhumed – an incredibly labour intensive undertaking.

Mr. Gao senior is an enthusiastic *garagiste*; wine equipment and barrels were scattered through various sheds, huts and go-downs. His small vineyard is being encroached by Yinchuan's suburbs; a line of high-rise buildings now looks down on the vines, which were just in the process of being resurrected. After sampling his Cabernet Sauvignon (distributed by Torres, very tasty) we drove out to see his new vineyard project on a huge, remote murrain plain filled with massive boulders. As he gestured broadly, drawing with his hands his vision of a huge new vineyard and winery, I could see the expression on my wife's face, which read "Here is someone even crazier than Chris".

Over the years I have visited a number of wineries around China. My earliest, and most remote, was on an investment trip to Urumqi in the far western region of Xinjiang. This domestically-listed company, called Summertime, had the worst balance sheet of any enterprise I have ever visited that was not yet in bankruptcy. The secret was that it was linked to a "*bing tuan*", or corps of retired soldiers, a system which the Communist Party has used to colonize the far west. I presume a general was sent to meetings with the bank manager. Summertime was involved in various activities but had at some point decided that wine looked promising. Being the army, big was clearly better and they had planted vast vineyards with huge factories to process the grapes. What the military men had overlooked, however, is that wine is not a simple commodity. To be sold to consumers, it needs branding, advertising and distribution. So Summertime now concentrates on shipping tanker-loads of raw wine to the big state-owned wine makers in the east.

The entrepreneurial Nanshan encountered a similar problem. I first met Chairman Sung Jianbo when he came into my office in Shanghai marketing his domestically-listed company. The company's main business is aluminium, but it also generates power, owns the largest supplier of men's suits to the United Kingdom, is the leading property developer along the Yantai coast, owns eight golf courses, mills flour and runs a hospital. At the end of this long recitation he mumbled "oh, and we also make wine – my worst business". Nanshan is only forty minutes drive from the castle so I went over to visit his winery. It is huge, with all the latest imported equipment in gleaming stainless. There are even two giant brandy stills (which, incidentally, we borrowed to turn our worst wine into wine spirit). But not a single grape. Wine is not aluminium, with a clear LME price for standard 99.99 grade. At the Sella & Mosca Winery, owned by the Campari Group, in Pingdu in central Shandong, they have also given over on grapes. The manager, Susan Zhang, told me that growing them was just too troublesome. The equipment is now only used to bottle wine imported in bulk. There may be an industry in China with a lower utilization rate relative to capacity, but I have yet to find it.

One of the pleasantest vineyards I have visited in China is that of Yunnan Red, which is invested by the big private equity firm, TPG. This vineyard is in the southwest of China, well below the "wine band" which supposedly delineates where wine can be grown around the Northern hemisphere. The secret is the elevation; 1,300 metres above sea level. The location is good from the point of view of sunshine days, high UV and a lack of bugs, but, like Shandong, it does have a problem with summer rain.

Unfortunately the vineyard is also mostly planted with edible grape types, which generally make for shallow wine. They have a back story about grapes imported and planted by a French priest, but I think a French priest would have known better.

I will end this chapter with the strange affair of our brief co-operation with Edward Davidson, a somewhat louche, but cheerful Englishmen who had pursued a chequered career which included, according to him, a period as Madonna's interior decorator. I first met him at the Wine Investment Council, the aim of which was to sell expensive wine even more expensively to rich Chinese and state-owned banks. We opened some samples and he told me that they might be interested to buy Treaty Port's entire output. This predictably never came to anything. Sometime later he turned up as the President of "The Imperial Club", apparently a swish watering hole in Shanghai. On the strength of a visit to Shandong, he decided that he wanted to stock our wines and whiskies. He talked excitedly about tall girls in short kilts selling Whisky Mc shots in night clubs. On the strength of this we organised two tasting evenings – one whisky, one wine – and I addressed a limited but well-oiled audience.

Our first shipment was on consignment. The plan was to count up how many he had sold and charge the club for them. When we duly tendered our first month's bill for RMB 9,696, we were told that Edward had fought with the lady boss and fled the country, and that we should ask him for our money. He turned up in Malaysia and said he had already given money to the club to cover the bill, so we should talk to them. My lawyers vainly sent the club a threatening letter. Robbed again.

This is the consolation email I received:

```
Jeez !!! how can they be so bad and cheat like
this !!!
```

Owner is Madame Chen and shes usually there
every day after 11am sitting in VIP dining room
on first floor you can tell if shes in as she has
a maroon mercedes parked outside

Sorry ;-(

Edward Davidson / Managing Director
Fine wine expert for Guopai auction house,
Shanghai , PRC
Advisor Shanghai international wine exchange

And finally my London taxi cab gave up the ghost in the same
week that Manganese Bronze went into bankruptcy. Local service
from their local partners, Geely, was terrible, so I was urged by
Big Che to pick a more prosaic MPV as its successor. But we got
240,000 km out of it, and some good advertising, so I guess I
should not complain too much.

INTERLUDE II

How we make wine at Treaty Port

To MAKE GOOD WINE you need good grapes. This is a truism, but my experience shows just how difficult this is, especially when you are a pioneer in a new wine area where the climate is not entirely sympathetic. For the big state-owned companies who dominate the Chinese market it has all been about boosting volumes to meet growing demand and grabbing market share. The market is not sophisticated. So, together with many local opportunists, they have tended to buy wine in bulk from all parts of China and overseas, then mix and bottle. Labels are not always accurate as to origin and vintage. This helps to explain why so much "Chinese" wine is poor. However, with growing wealth and travel, the wine drinkers of China are starting to understand what good wine should taste like, and which wines best suit Chinese dishes. This is putting pressure on the larger companies to improve their act. With so many food-related scandals, there is a desire amongst consumers for the authentic. This is allowing the development of small estate wineries, offering premium wines with their own character, such as at Treaty Port.

We are very careful about the sprays we put on our vineyard, and for the last month before harvest, we do not spray any chemicals at all. Our earliest harvest has been in August, our latest towards the end of October. To assess ripeness, we measure sugar and flavour. For most grape types we are aiming at sugar levels of 23 to 24 brix, meaning about a 13.0 percent alcohol in the finished wine. This may only be achieved in the best varieties in better years, however. We watch the weather forecast for signs of rain. My experience is that Chinese farmers tend to pick too early, keen to get the harvest in before mildew can ruin the crop. The resultant wines have a neutral or a "green" under-ripe flavour.

At Treaty Port we handpick the grapes and the maximum we can harvest in one day is about seven tons of grapes. The average yield is about 300kgs per *mu*, so this is about 20 to 25 *mu* picked in a day. This is employing 40 people, and starting at the crack of dawn when it is still cool. We pick using plastic baskets which hold 25kgs each.

These are stacked on a cart and delivered to the winery by tractor. We pick each grape variety as it becomes ripe, the whites first, starting with Muscat a petit grains. For the reds, Merlot ripens the earliest, and Petit Verdot and Cabernet Sauvignon are usually the last to come in.

The grapes are sorted twice – a benefit of hand-picking over mechanical harvesting; once in the vineyard and once on a conveyor belt at the cellar. Unripe grapes, bugs, leaves and sticks are removed. The bunches then fall into a machine which de-stems and crushes the fruit, and this crushed grape mixture ("must") is pumped into our 3-tonne screen press. Red must is pumped directly into the stainless steel fermenters where it soaks before fermenting for less than two weeks. We have fourteen large tanks, made locally to the design of our Australian consultant winemaker, Mark Davidson, who has assisted in the design of the winery from an early stage. Each fermenter is suitable for whites, rose or red wine production and is divided into an upper section for red fermentation and a lower for draining off the upper tank or for wine storage. Both tank sections have their own cooling jacket to manage fermentation and storage temperatures. The must is inoculated with a suitable dried yeast strain to start the ferment. The temperature is controlled at 25C° to 30C° for reds and below 15C° for whites and rose. During red wine fermentation, skins, which float to the top of the tanks, are wet down on a regular daily program either by irrigating the juice at the bottom across the surface of the must through in-place pumps on each tank or by draining juice from the top compartment to the bottom section which is then pumped back over the top of the ferment. This process is known as "rack and return" and assists the oxygenation, colour extraction and cooling of the ferment. In addition to the large tanks, we have six variable capacity tanks which are used for smaller batches of fruit. We do not add sugar ("chapitalize") to the ferment. Instead of pressing the red skins after the ferment we use the draining screen in the top tank to separate the juice from the skins and thereby reduce the more coarse tannins in the final wine. This means however that our average "free run" wine yield from the grapes is a little lower – 650 litres per tonne at best.

Reds and chardonnay, in contrast to Muscat and Viognier, undergo a second "malolactic fermentation", which, induced with bacteria, softens the acidity of the final wine. The design of our winery, with a very high ceiling and thick walls, keeps it cool in summer. One drawback, however, is that by the time we get round to malolactic fermentation, it is getting cold in Shandong, so we typically have to use heaters or an electric blanket to make the wine warm enough (above 15C°) to successfully complete the process. This usually takes about one month, and we are pleased if we can finish it before Christmas.

For the better quality reds we have been maturing them in 300L hogshead barrels for about a year. We have experimented with various types of oak and find French medium toasted seems to work best for the reds and Hungarian for the Chardonnay. Muscat does not see any oak, but some Viognier has been stored in old barrels for texture.

For the 2014 we have started to experiment with a new Australian technology called "Flexcubes". These tanks, which come in 1,000L and 2,000L sizes, are made of polymer and "breathe" like a barrel. Oak staves (the same as used in making barrels) can be introduced as required. The benefit of this technology is greater control over oak content and an ability to get fine wines ready for bottling earlier. Whether we use barrels or Flexcubes, these must be kept topped up to prevent oxidization of the wine.

Sulphur also helps against oxidization and to end the second fermentation. Before blending and bottling the wines must be fined - we use casein, a milk protein, in whites and rose, and egg white in reds. After cross filtering (we use a Pall filter from Germany) we are ready for bottling.

Our Italian bottling line is over engineered for purpose – we could bottle our whole production in a few days. The problem is that in China you cannot get a wine production license without owning a bottling line, so we cannot outsource bottling, as happens in other countries. There are therefore a lot of under-utilized bottling lines in China. I decided to use the screw cap as a bottle closure, rather than cork and capsule.

This stems from my frustrations with corked wine (after someone's gone to all the trouble!) and the difficulty of getting good quality corks in China, not to mention the low distribution of corkscrews here. After an initial problem with alignment, which made the screw-caps difficult to open, this has worked out well. We have had no problem sourcing good quality packaging materials locally. However, with limited space, we need material deliveries to be just-in-time.

We have a laboratory equipped with all the basic equipment required. It sits on top of the granite cliff, which forms the Northern wall of the winery. Despite the best efforts of the Hygiene Bureau, I managed to prevent this from being covered in concrete. I think it well illustrates our "*terroir*". In rainy weather, a small stream runs down the cliff and through the winery.

登高望遠
龍騰虎躍
紅紅火火
酒香情深

CHAPTER 12

THE FRENCH CONNECTION

A man should swallow a toad every morning to be sure of not
meeting anything more revolting in the day ahead.
Chamfort

WE HAVE NOW reached December 2013. This chapter was written
from my office in Shanghai. Down below, the busy Huangpu
River glittered in the winter morning sunshine. The towers on
the other side of the river, however, are mere ghosts wrapped
in thick smog. The previous week, pollution in Shanghai was
the worst ever recorded – the PM2.5 count was over 600, eight
times the level considered unhealthy. The suppliers of masks and
air purifiers are doing a roaring business. At such a time, it is a
pleasure to journey up to the vineyard for some greenery and
fresh air.

When we started this project, no Chinese in his right mind
would voluntarily visit the countryside and, if you were
unfortunate enough to be born there, the key imperative was
to escape as soon as possible to the excitements of the big city.
I remember one official on a visit to our vineyard staring with
disgust at some mud which had stained his black shoes. I have

started to notice a change. Individual tourists now call in at the castle, proudly displaying their shiny new cars. The local villages, Mulangou in particular, have been beautified to receive them, with whitewash, street lighting and miniature parks. The introduction of a twice-weekly rubbish collection service has been a big help, though the farmers still happily discard the bags used to protect apples where they fall, as well as the silver foil used to tan the apples, leaving them to blow around the vineyard. We hired an assistant wine maker, Hans Zheng, a Qingdao boy who, having returned from working in Australia and New Zealand, professed his desire to live in the peace and quiet of the country (though I think his girlfriend might have other ideas).

One of the innovations in Mulangou is a signpost pointing out directions to the local sights. One says "Frenchman's Cottage", this being the house renovated by Gerard Colin. He no longer works for Lafite, but his replacement, Olivier Richaud, has proven friendly and co-operative:

Le 29 avr. 2013 à 11:18, "Olivier Richaud" <orichaud@dbr-citic.com> a écrit :

Dear Chris,

It was a real pleasure to make your acquaintance yesterday.

Once again thank you very much for the great lunch and the wines ! Sea food and Pol Roger champagne were delicious!
Christophe, Eric and I really appreciate that you offered to open your winery for us having

```
our vinification during the coming harvest.
By then, we'll have to discuss the practical
and financial conditions.

I do confirm too, that if you still agree,
our trainee who shall work on diseases
identification, could make some kind of
comparative study between our two vineyards.
Looking forward to meeting you soon.


Regards,
Olivier
```

Lafite started to construct their winery. The photograph on the billboard placed out front showed a low, modernist structure hugging the contours of the hill – very different I am sure from the grand chateau which their JV partner, CITIC, might have wished for. Still it will, apparently, be the only Rothschild property in the world to have a shop. I also saw that it would include the Baron's private residence. The winery was not finished for the harvest (and they will do well to finish it for next year's). So Lafite's first ever vintage in Asia was made in our cellar! There were only 1.7 tons of grapes, but the longest journey etc. We flew the tricolour over the castle in Olivier's honour for the arrival of their grapes. Lafite's harvest suffered from some of the same problems as our own, particularly their Marselan and Syrah. Hopefully their researchers can come up with a solution. However, Olivier seems committed to growing the grape bunches as close to the ground as possible, with no grass cover, the opposite of Mark's strategy.

In addition to the small Lafite batch, we processed Riesling for Luye, as last year, but this time with a properly calculated price (about RMB 11,390 per ton of wine). They left their red grapes out too long, so we made no red wine for them. Mark's advisory contract with Luye came to an abrupt end during the summer as the nature of the project became evident. Over a coffee, the owner Mr. Liu told me he had already sold nine chateaux at RMB 30 million each and hoped to complete ten more by the year-end.

We also processed 4.7t of red grapes for K.C. Kung's Xiandao vineyard; the quality of their grapes, especially Merlot, was better than ours.

The revenue from toll processing helped to offset a fall in wine sales. The new president Xi Jinping's crackdown on conspicuous consumption by the Communist Party has hurt all luxury goods producers in China, who are heavily dependent on the gift-giving tradition. Our over-reliance on sales to local government therefore cost us dear. Wines sales this year fell sharply. The number of tour parties visiting the castle increased, but our benefit from this is limited to a small entrance/tasting fee. They buy little wine to take away; our price points are too high for the average Chinese tourist. We need to design something below RMB 100 to meet this demand.

Local studios started to pay us to take wedding photos at the castle, a sine qua non of any wedding in China, but this has not yet translated into the next logical step of full wedding bookings, probably because we are too remote, making transportation of guests difficult. We also spent money re-designing the website and opening a shop on the virtual mall Taobao but did not sell a single bottle online. Total revenue this year, including accommodation and food, was just RMB 1.3 million, down 37 percent year-on-year, for an operating loss of RMB 640,000, similar to last year.

This year we managed another self-inflicted disaster even

before the season began. I discovered that our aromatic "Lady of Fashion" white had undergone secondary fermentation in the bottle, causing it to become fizzy and lose most of its taste and aroma. The same happened to the Rosé and Luye's Riesling. Our inquest looked at quality of water used for cleaning the bottling line, the quality of food-grade gas used in bottling, the yeast and the cold-stabilisation period, but no specific smoking gun was found. We just have to be more careful at every step. We re-bottled the Luye white (on which we were already losing money) and held our own white and Rosé to blend into 2013's Rosé. The only new bottling we did this year so far was the Castle Red 2011, which included an addition of Mark's Australian Merlot to give it body.

The winter and spring was very dry. On May 18th I sent the following email:

Mark,

I met up with the mayor yesterday, who wants me to allow them to take the end off a Marselan plot (about 50 vines) so they can build a straight path to the door of the over-sized new temple that's been built. My arguments about the perfectly good road around the side of the vineyard, and the fact that, whilst communist and Confucian roads are straight, Taoist roads are inevitably winding, bore no fruit. I conceded (as I had to) in return for a promise that we can collect this harvest, they will relocate the vines in a new piece of land and will help with the castle deeds.

Whilst all this was going on, I got chatting with the Taoist monks. I told them that the spring had been dry, so if they could pray for some rain I would appreciate it. That evening, out of a cloudless sky, persistent rain. Brilliant! Much better than drip irrigation – it only cost us a packet of cigarettes. I have asked Emma to work the Taoists into our August preventative spraying program. I mean, it's organic, and how many vineyards are blessed by Taoists?

Actually, it's raining so hard now it is starting to come through some faults in the roof, so I will ask them to turn it off on my way out…

Regards
Chris

What happens next has the inevitability of a Greek tragedy. Rainfall in July 2013 was ten times the average. It rained so much that parts of the vineyard started to resemble the Amazon. Still the improved spraying regime kept the vineyard looking healthy.

The leaves remained green and mould-free. Even a few weeks before the harvest Mark was predicting a 50-ton harvest. This time it was the grapes which rapidly shrivelled, leaving us with acres of dried-out grapes. The final harvest was a pathetic 6.9 tonnes, of which 0.5 tons was white grapes. Again the reason is unclear – perhaps a failure to spray properly at the time of flowering? Sometimes I wonder whether Shandong is a suitable place for growing grapes. Perhaps there is no such place in China, despite its size. In the east it is summer rain; in the west it is the freezing winters. As David Foster Wallace said "The truth will set you free. But not until it has finished with you."

We tried out a spray machine that autumn. It was not just a question of automation in the face of increasing wages, but also reflects a desire for a more even spray than can be achieved by hand. We planned to buy more machines for the new season, so in the winter we brought in the end posts where necessary to allow a turning circle at row ends. The time had probably come to cut down some of the more unsuccessful varietals (Cot, Sangiovese, Viognier, Arinanoa) and replace them with more hopeful ones (Marselan, Merlot, Muscat). It is a tough decision to acknowledge that all the effort we have so far lavished on these plots was wasted.

The government sent round a digger to plough a straight path through the Marselan plot below the temple, but Emma sent it away. We still do not have our deeds, though we are now surely in the final stretch. The fire department finally deigned to pay attention to the castle four years after its completion. We have been forced to add rope ladders to the various balconies. This has the strange effect of making it look more medieval ("Can I lower you to the ground, my liege?"). I have been "asked" to make a donation to the local fireman's club. After a grand opening on Qiu's birthday at the start of the year, the temple itself remains unfinished. Apparently the priests have left in disgust.

Our long-serving chef, Cai Hong, unfortunately decided to leave to help her brother's business in Hainan (taking tourists on fishing trips). Emma has not been able to replace her, so the quality of food slipped. Also, in the interests of cost saving, little repair work was done on the castle, so slates continue to slide off the roof. The only additions to the castle this year were a Feng Zikai painting and a modern sundial from Scotland erected in the herb garden.

The sundial is inscribed with two lines of a poem by Mei Yaochen (1002 – 1060) from the Sung dynasty, which seemed appropriate:

菡萏花迎金板舫，葡萄酒瀉玉壺漿

To Mr Xu on being posted as Yuezhou Magistrate

Glad to hear you are posted to Yuezhou; in May the peaceful lake is like a mirror.

Lotus welcomes the golden boat carrying you, grape wine flows from the jade cup.

The hills are quiet and the clouds blow west; the spring is cold after the rain.

The emperor cannot bother you here; reciting poems is a good way to keep healthy.

In like vein, my Open Door team from Shanghai made an outing to this distant rural spot. One of the fun parts of the project is being able to show it off, to see the (usually) amazed reaction of visitors. The team, with partners and children, played the part of tourists to the full, visiting Penglai Pavilion, riding horses, hiking up Qiushan and eating and drinking heartily.

Amongst all the trials and tribulations of building a Scottish Castle in China, and trying to make a drinkable wine, one thing which has sustained me is the reaction of guests. I still get a kick out of the widened eyes, and open mouths, as visitors get shown into the Great Hall, or the surprised nods from wine aficionados as they realize that actually our Chinese wine tastes pretty good. From foreigners the most common comment, especially when walking on the mountain behind the castle is "This does not feel like China at all". From the Chinese visitors, there is a great curiosity about where things come from, how much they cost and who I am ("Is the owner an aristocrat?"). Many of these comments are captured in the series of large leather guest books placed at the reception desk until full, and then moved to the library.

There are comments in a surprisingly wide range of languages. Chinese is naturally the most popular, followed by English, but there are also quite a few comments in Korean (Yantai has quite a large Korean community – we are not far, as the missile flies, from Pyongyang). The next most frequent are German and French, with the odd comment in Spanish and Italian. We even have one comment in Greek, written by a gentleman glorying in the name of Phaedon Tamvakakis, who translated all the novels of John Fowles into Greek.

There are the back-handed compliments: "fantastic folly", "this wine is almost as good as that from France" (I translate from the French). There are some rather misty-eyed comments from travelling Scots "A little piece of Scotland in China. For two Scots, the most amazing experience". There are more professional commendations: "The service, food and accommodation were all wonderful. A truly memorable experience." This from Paul Mooney of the South China Morning Post. A personal favourite, from a nameless Chinese visitor is: "So impressive! I think the boss must be very romantic!" This last comment is accompanied by a Japanese-style cartoon, which unfortunately looks nothing like me. The books contains a lot of drawings of castles and princes/princesses by budding young artists.

One of the most exuberant comments comes from a Latin American businessman, a customer of my financial business, which starts *"Hola Amigos!"*. I had given him workers' gloves and told him he needed to earn board and lodging by green-stripping – the back-breaking task of pulling shoots off the lower part of the vine. He took it in good heart. I don't think many vineyards can claim they have been tended by Mexican multi-millionaires.

One point to note on the Chinese comments is how many of them are in poetic form. Whereas in the West, poetry has become a distinctly niche interest, here in China it is widely taught. Many Chinese can recite by heart a range of famous poems from the form's apogee in the Tang dynasty. The poems on pages 90, 104, 128 and 149 are the best original compositions taken from my reception book.

INTERLUDE III

Art at the Scottish Castle in Shandong

ONE OF THE MOST enjoyable parts of this project has been the creation and collection of art for the castle. My aim was to create a beautiful and distinctive interior; so many vast and expensive edifices erected in China have disappointingly garish decoration.

Robert Koenig, from Manchester but of Polish extraction, was a veteran of my Dairsie campaign. He carved the beautiful entrance doors and double doors to the Great Hall. His brief was to use a mixture of Scottish, Chinese and vineyard themes. Robert Koenig also carved the statue of the Scottish adventurer, arriving on his ship, which greets visitors to the tasting room, as well as the tasting room "grape table" and the Polaris Seat (you use it on a summer's night to find the North Star; the stars are bright over Qiushan).

The "Scottish Adventurer" history derives from a painting by Gheerraerts the Younger from the 1590s, which hangs in the Great Hall. The identity of the gentleman is unknown, though the hand resting on a helmet indicates pretensions to military accomplishment. With his luxuriant red beard, he could easily be Scottish, and Gheeraerts the Younger painted a number of men from the circle of the Earl of Essex (also red-bearded). So I have been developing a story whereby the gentleman, let's call him Sir James, becomes implicated in Essex's disastrous rebellion against Queen Elizabeth, is forced to flee and, after many adventures, is caught in a typhoon and washed up on the Shandong shore to be met by Ming officials. He helps them fight against the invading Manchus. This explains, of course, why there is a Scottish castle in the middle of the Chinese countryside. I go on to explain to increasingly wide-eyed party officials that this contributes to the relate occasional appearance in the Shandong population of taller men with reddish hair. And on his return to Scotland, he bought back the Chinese recipe for blood pudding, now known locally as haggis…

We are clear about the identity of sitter in the other early British painting hanging in the Great Hall. Michiel van Miereveld painted this portrait of Sir Henry Wotton in 1620. Sir Henry, a friend of the poet George Herbert, was ambassador to Venice. It was him that defined an ambassador as "an honest man sent to lie abroad for the good of his country".

It was Robert Koenig who introduced me to Roman Flezar who, with his wife Ewe, is responsible for the stair murals and the painted ceilings in the Great Hall and in the entrance pend (look up to see a portrait of Eliza amongst the swallows). All were executed under difficult conditions. Roman's large mural on the main stair shows Sir James' departure from St. Andrews and arrival in Penglai. Amongst the well-wishers seeing him off you might spot the faces of the three wine-makers and myself.

Amongst the group of Ming officials greeting Sir James there are the faces of Mayor Liu, Mayor Zhang, Tiffany, Mr. Hao and the terrible Mrs. Ren and her husband.

Renny Tait, an Edinburgh-based artist with a speciality in architectural paintings, had painted a picture for me of Dairsie

Castle. For the new project, I commissioned a series of six paintings of interesting Scottish castles. We have already used a number of these as labels for our house wines – Castle Red and Castle White. I later added an additional painting in the same style of the local Penglai Pavilion, which Renny, being an inveterate non-traveller, painted from picture postcards.

One Scottish painter with three paintings in the castle is James Morrison. I think he catches the Scottish sky so well.

One of my favourite modern Chinese artists is Feng Zikai (1898 – 1975). I bought this "Harvest Year" at a Bonhams auction in Hong Kong, having previously unsuccessfully bid for a couple of his other works. I thought the subject matter suitable for a vineyard. Just after this purchase, the artist's pictures, including this one, were used extensively in a nationwide poster campaign by the Communist Party with the theme of the China Dream.

The 18th century Flemish tapestry and 17th century bust of Bacchus in the living room also come from Bonham auctions, as do a number of the pictures in the bedrooms.

The antique furniture comes from Georgian Antiques of Edinburgh and Andy Thornhill, a reclamation company in Brighouse, West Yorkshire. Just to show that I don't just collect ancient or figurative art, our tasting room has two abstracts by Seattle artist Gregg Robinson, one entitled "Perfectly Normal", not a statement which could ever be applied to the Scottish Castle in Shandong.

CHAPTER 13

The End

Under capitalism, man exploits man, while under
communism it is the other way around.
Old Czech joke

The essence of authority is that it must manifest its power.
Ryszard Kapuscinski:

THIS IS NOT THE final chapter that I wanted to write. That chapter
would end with a triumph; after overcoming so many adversities,
Treaty Port Vineyards would enjoy a fine, full harvest and produce
a vintage so good as to force the wine experts to revise their
views on what is possible in China. A happy, local population
would value the prosperity bought to this rural backwater by
the project. A settled and competent local management team
would be able to develop the business independently, allowing
me to relax and enjoy my visits to the vineyard. The broad, sunlit
upland meadows would be in sight.

Instead, I discovered in March that the government intends to
build an elevated, four-lane highway through the middle of my
vineyard, in between the castle and the lake. This is not linked

the new Yantai-Penglai international airport, which is to open shortly, but will be the main North-South motorway taking traffic from Manchuria to Qingdao should the highly ambitious 60km tunnel and bridge across the Bohai Gulf from Dalian to Penglai ever be built. This is doubtless the same motorway that we were assured in two years ago would go the other side of the lake.

The first map I saw was sent to me by Lafite. A light green line shows the final route chosen, threading between the lake and Qiushan. The new route, about which there was no consultation, ignores the possibility of utilizing the old route to Qingdao, and just widening it. There were also marks in red showing where the new airport would be. Yes, we were on the flight path. There was another map which showed how we think the road would affect the vineyard (the numbers in red are plot numbers). A couple of hundred metres to the East and they could have avoided the vineyard entirely – as it is, the motorway goes straight through the best land, where much of the Marselan is now planted.

I sent the following letter to every politician that I know, even the British ambassador, quoting Premier Li Keqiang's recent speech on the "the war against pollution".

Replies received I none. I called my journalist friends, but they told me land seizures are so common it is not even a story.

Mr. Sun, the new party secretary in Daxindian, my fourth since the project began, was apologetic but assured me that the route has been fixed by higher echelons and is set in stone. There was nothing he could do about it. He lolled back on the sofa in his office on a Sunday morning, relaxed, smoking thin, gold-tipped cigarettes and taking sips from his jar of tea. The only time he looked concerned was when I told him I could help him to garner some attention, and with that Emma unrolled a large, red banner we had prepared which read, in Chinese "Protect the

Environment! Oppose the Highway!"

The cigarette stopped half way to his mouth. He leaned forward.

"I wouldn't do that," he said. "Some farmers may follow you. I might have to call in the police. People might get hurt."

He held my stare for a few seconds, before leaning back and reassuring me that work would not start until after the apple harvest. After nine years of investment, I was compensated at the standard rate for "grapes over three years" (which I am later informed is just RMB 18,000 per *mu*). The road looks as though it will occupy about 33 *mu* of the vineyard (about twelve percent) directly, but the indirect effect of shadow and blocking of the important summer breeze are impossible to predict. I have still never been shown a detailed plan with elevations. There will also be two years of over-filled trucks, dust and construction worker debris.

As if to get in on the act, the Daxindian government, without consultation, decided to build its own new, road right up to our door. It was completed in a couple of weeks. It was a "tourist route", linking Aishan hot springs, a horse-racing track, Lafite and the Scottish Castle, to whisk bus-loads of tourists conveniently between one and the other.

I wondered aloud why tourists would want to come and see countryside covered in cement and noisy with traffic, but Mr. Sun looked at me uncomprehendingly. He had received a grant from the tourist authority and would spend it (and accrue the associated benefits) before it went away. The incentives for local party officials are still heavily skewed to short-term economic gain: who cares about long-term effects to the community as the official will be on to his next assignment before these become evident? It reminded me of my first-ever job as a brand assistant at Procter & Gamble; the real effect of the advertising campaign was not evident until after the brand manager had been promoted away, so the key was, as Machiavelli said "Videri Quam Esse" (to appear to be, rather than to be.) Environmental protection, which only shows effects in the long-term, and can depress economic numbers in the short, is therefore merely another slogan; a gesture like the unconnected windmills that make the surrounding hills resemble an enraged porcupine.

Objections from the great Lafite, backed by the powerful CITIC Group, proved completely useless, even with the officials of this tiny town. Ten *mu* of Lafite's vines, planted only in 2011, were dug up, together with a forest of apple trees. On the road back from my meeting in the town hall I passed a yard piled high with trunks, gnarled roots and an industrial scale wood-chipper. Two of our ponds were built over without my permission (RMB 20,000 compensation was promised.) The straight path to the door of the Qiu Temple had not been forgotten and a parcel of Marselan vines had to be moved to make way for it. The temple itself stands unfinished, one wall unpainted; the monks have departed and the evergreens lining the wall have turned brown for lack of care. So municipal funds will be spent building an unnecessary path to a locked temple; it is difficult to beat that as a metaphor of what is wrong in China.

It looks at least as though Pengtai's horrible motel plan has been abandoned due to a lack of finance. This is a mercy, but comes only after the company hugely expanded the quarry, scarring the hillside. Instead, the Leung family from Singapore, that owns a company called Metrotown, has decided to build "the Runaway Cow" resort in between the castle and the village of Mulangou. Work has already started on the first stage of a radical German-designed hotel. At least the Leung's are friendly and pay attention to design (I have already met them with

Hungarian and German architects). They do, however, seem terribly optimistic (a bit like me on page 1); they think the resort will be finished next year. I had imagined that for a company specializing in resorts, the accompanying vineyard would be a token gesture, and we could help them with it. But no, despite all our war stories, the vineyard is to be 400 *mu* and, their adviser being from Burgundy, it is to be planted with that most difficult, thin-skinned grape, Pinot Noir!

There is just one more bizarre episode to include before I close this episode. Through my link with the stove maker AGA, I had made the acquaintance of "brand ambassador" and celebrity chef James McIntosh. He was making a TV programme in China, so we arranged for him to visit the castle. My idea was that we would cook a dinner for a VIP audience and help train our staff on cooking with the AGA. I sent out the invitation.

Unfortunately it turned out that his visit was so brief, he was only able to cook a pear dessert, which did not prove terribly popular. But at least he had a good time, as can be seen from the following email. He was accompanied by his German boyfriend Thomas. I was unable to attend, as I was back in Yorkshire to be with my mother in her final illness.

```
Chris

I've been very mindful tonight about your family
situation during a very high energy dinner at
your castle.

Firstly wow does not describe this family testament
you have created.

It's way off the scale at the end of truly
```

magnificent.

Many in Beijing in the global wine world were told about your endeavours. Janis gave me a copy of his book and it's now in your library.

When we arrived we laughed so much at the taxi. A fabulous touch. Emma was away and the girls did not understand that Thomas and I would like the same room. We laughed so hard at it so they gave us the bridal suite. It was so funny!

A TV crew were here for the wine from a Shandon station. They were showing the show in China and Germany. Thomas who has done no TV in his life did one version in German (he's from Frankfurt) and I did another in English. However the make-up lady drew new hair onto Thomas bald patch. We did laugh.

Tonight I cooked the famous Scottish dessert of cranachan. Only I did it Chinese style. I will forward the recipe later as it could be a treaty port special.

Thomas' dad is called Hans, so yet more laughs. In the am we are seeing how Hans makes the wine and then off to Hong Kong for a few days.

I'm rather blown away by this place. I'm thinking of talking to China Food TV who are based in Quingdao about doing a James at home show here in the castle. What do you think?

A fabulous day so far. Emma is fantastic. I will
do some AGA training in the am.

Looking forward to hearing from you soon.
James and Thomas

饕餮盛宴 海鲜大餐 Seafood feast

2014年5月26日 晚18：30-21：00
May 26th, 2014   18:30pm-21:00pm

　　来自英国的超级厨师James McIntosh将结合当地特色海鲜，搭配登龙酒庄自酿高端葡萄酒，为您呈现一场美轮美奂的饕餮盛宴，期待你们的光临！

Super Chef James McIntosh, Britain's food ambassador and award
winning cookery book writer, will serve a unique meal combining
local seafood specialities, British flair and Treaty Port wines.

258元/位 258Rmb/P.P

Email:sales@treatyport.com
Website:www.treatyport.com

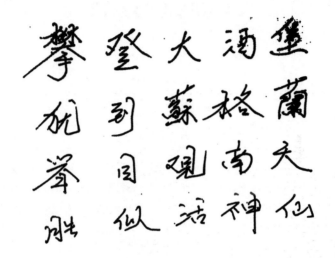

堡蘭山仙
酒格南神
大蘇觀活
登到目似
攀猶舉勝

INTERLUDE IV

THE WINE INDUSTRY IN CHINA

ARTICLES IN THE Western media about the wine market in China do not, I have found, tend to be profound. They mention the old canard about Chinese putting Sprite into their wine. I have never seen this happen myself, but China is a big place, so I am sure that it happened somewhere (and some of the Chinese wines I have tasted over the years would be improved by Sprite). The foreign media likes to focus on the obsession with Bordeaux Premier Cru. Certainly a lot of Bordeaux is drunk in China (quite a lot more than is sold, by all accounts). But this is largely a function of a lack of knowledge about what is available; how can a Chinese host be criticised if he serves a world-famous brand, even if it does not necessarily pair with the food served? But as knowledge grows, and the Chinese consumers I have met are keen to learn, I am sure this will change, and the selection of wines will become more self-assured.

Chinese consumers will start to drink what they like, and not what they think they ought to like. I have often seen consumers served a flinty, tannic wine from a bottle with a famous label and say "Oh, very nice", whilst from their forced smile you can see that inside they are saying "*suan*" (tart).

The main point of most articles nowadays is how big the Chinese wine market is, how it has just overtaken this country in total wine consumption or that country in terms of wine production. Large numbers are thrown around that are difficult grasp and not terribly helpful. For example, according to the International Vine and Wine Organization China in 2014 China produced 11.8 million hectolitres of wine and consumed 15.8 million hectolitres. (There, I told you.)

These large numbers are not surprising, given the scale of the population; anything multiplied by 1.3 billion is a large number. When the figures are brought back to a per capita basis then the number is usually less impressive. The Chinese wine consumption of just over one litre per person per year compares poorly to that of 15 litres in the U.S., never mind the 50 litres in France. Chinese wine consumption has increased rapidly in the 2000's, albeit from a low base; OVI estimates 136 percent growth between 2008 and 2013, in contrast to the 18 percent slump in France. However, the wine business remains a minnow relative to the local spirits and beer markets; wine is estimated in 2014 to represent just three percent of all alcohol consumed.

In 2013 the rapid growth period came to a sudden halt, when consumption fell by 2.5 percent, the first fall in 10 years. The decline coincides exactly with the accession to power of Xi Jinping and his crackdown on party corruption. The slump in government orders in 2013, which I witnessed first-hand at Treaty Port, was reflected not only in the economic performance of other wine makers, but in the performance of almost all luxury items. This illustrates the important role which gift-giving plays in Chinese society in general, and the wine business in particular. At Treaty Port we do a good business in RMB 20 wood boxes into which customers can place the wine they have purchased, to add to the perceived value of the gift.

The distribution of wine in China has a different character from that in the West. Specialist wine stores seem to be a very bad business judging by the rapidity with which they open and close in Shanghai. On the whole, the Chinese still do not buy wine to consume at home. Wine is drunk in restaurants and night clubs.

Most of the wine drunk is red (approximately 90 percent) for historical and cultural reasons (lucky colour). But I am sure that this will change as restaurants have more capability to chill whites, and consumers get to know how well whites can pair with many Chinese dishes. In tastings I have found that fruitier white wines with a nice floral bouquet go down well, but this has not yet translated into wider sales.

One of the major challenges to developing an appreciation of grape wine in China is the "ganbei culture". In restaurants and clubs alcohol is not sipped, but thrown back in one go. More than just a social lubricant, it becomes a macho ritual; the men (and it is always men) demonstrate their virility through their ability to hold alcohol. The noise level escalates quickly. Now this is probably the best way to drink most local spirits, but it is a terrible waste of a good wine. Even in more sedate gatherings, it can feel awkward to take a sip without toasting someone else at the table.

Another major obstacle to developing knowledge of wine is the unimaginative way in which it is sold. How is a Chinese consumer to choose a wine in a restaurant when it is just a long list of strange foreign names? This is intimidating; far better to have a shorter list properly explained, with suggestions as to which of the restaurant's dishes are best suited. Shoppers in supermarkets are left to choose their wine merely by label design and price. Even in specialist wine shops, often little is done beyond grouping wines into country of origin. Once when interviewing the owner of a dental hospital I noticed, beneath the glass-topped coffee table, a yellow-covered "Wine for Dummies" book, translated into Chinese. The thirst for wine education is unslakable.

CHAPTER 14

Not quite the End

What wondrous life is this I lead
Ripe apples drop about my head;
The luscious clusters of the vine
Upon my mouth do crush their wine.
Andrew Marvell

IRONICALLY, 2014 TURNED out to be by far our best harvest. This was in part due to the dry summer, but also to better management by Hans and his ex-colleague Shawn, who we hired once Huadong abandoned its vineyard at Qixia. It should be noted that our neighbours did not enjoy the same bumper harvest as us. The spray programme was pro-active, with particular attention around the crucial flowering stage. Weedkiller was used under vines and the strips in between mowed. We started picking white grapes on September 2nd, but did not pick our last Cabernet Sauvignon until October 28th. As an experiment we bagged some of the white grapes, just as the local farmers put their apples in brown, waxed bags. This proved successful, allowing us to reduce chemical use whilst still protecting from insect damage.

In retrospect we should have done some green pruning, but you can understand our reluctance given the disappointing harvests so far. In the end we harvested 102 tons of grapes, with good flavour and sugar content mostly in the 22 to 23 Brix range. This compares to just 37 tons in our previous "best" year, 2009.

Variety（品种）		Picking Date	Mu （亩）	Quanitity 数量 （Ton 吨）	Output (Kg/mu)	Brix	PH
小白玫瑰	Muscat	01/09/2014	14.67	2.883	196.52	19.3	3.37
霞多丽	Chardonnay	01/09/2014	11.5	2.783	242.00	21.4	3.46
维欧妮	Viognier	01/09/2014	13.99	2.15	153.68	22.7	3.54
西拉	Shiraz	13/09/2014	40.85	12.543	307.05	22	3.52
高特	Cot	13/09/2014	8.63	0.97	112.40	mix in	shiraz
梅鹿辄	Merlot	26/09/2014	33.05	15.474	468.20	23	3.53
桑娇维塞	Sangiovese	27/09/2014	8.59	4.213	490.45	21	3.53
黑歌海娜	Grenache	27/09/2014	25.4	7.94	312.60	22	3.48
品丽珠	Cabernet Franc	09/10/2014	13.4	9.511	709.78	22.8	3.59
阿娜	Ana	11/10/2014	7.91	3.94	498.10	mix in	Petit Verdot
小味儿多	Petit Verdot	Oct 11/29	16.14	7.8375	485.59	21.4	3.51
马瑟兰	Marselan	Oct 23/24	37.19	15.35	412.75	24	3.48
赤霞珠	Cabernet Sauvignon	Oct 18/29	44.45	17.167	386.21	24	3.52
Total合计			275.77	102.76	372.63		

We did not do much toll production this year, only making some white wine for a vineyard in Zhaoyuan with a Japanese advisor, as we were undercut by a factory in Penglai. But given our large harvest, this was probably fortunate, and we would have struggled to cope with much more. As the year ended, equipment was delivered for us to start cider production in

March 2015 for a brand called "Cider Republic" on an OEM basis. The scale will not be large – the customer wishes to supply 30-litre one-way kegs to expat pubs, exploiting the 70 percent import tariff on imported cider – but should help offset some overheads.

So as I write this, on the 10th anniversary of starting this venture, I now have 70 tons of wine down in the cellar and need to upgrade my sales effort. I have cut the retail price of our castle Red 2011 and Rosé 2013 to help move them out to make way for our 2014 wines, which promise to be more than decent. There is an elegant Chardonnay/Viognier blend that we are calling "The Lady of Fashion" and, my personal favourite, a rich Petit Verdot/Arinanoa blend which I think I am going to call "The Prince" after Manchu Prince Gong, who took a progressive line in dealing with foreign countries, and of whom I have found an excellent old photograph by Scottish photographer John Thomson. We have already sold some of the 2014 *en primeur*, in barrels. But you never know, there may still be a little left by the time you read this...

POSTSCRIPT

As I WRITE THIS POSTSCRIPT, we have just started to pick the first of the red grapes, Syrah, for the 2015 vintage. The white grapes are already safely gathered in. I don't want to jinx things (I now understand why farmers are notoriously superstitious) but we seem to be finding a way to make decent wine in Shandong. The season was not especially favourable – the spring was too dry and August was wet – but through better management, and a stable, professional team, we seem to have nursed the crop through to the fine, long autumn, which is typical in these parts. This year we might be able once again to leave our slowest-to-ripen varieties (Petit Verdot and Cabernet Sauvignon) on the vines until the second half of October. My challenge now is to try to keep the team in place. I hope that this will become easier as more estate wineries develop in China, local knowledge of wine-making improves and Qiushan becomes a wine destination, helped by completion of my neighbour Lafite's winery in 2016.

Now that we have a more consistent product, I need to improve our marketing. Given our small size, traditional advertising is unaffordable. However, we are setting up a WeChat shop (WeChat is the most popular messaging service in China) and I plan to use this as a platform for some "social marketing", helping to explain wine and our products to those in China with an interest. We also need a good distributor to represent us in Beijing and Shanghai.

Our hotel operation will only ever be a small part of the business; we only have six rooms in the castle and six in the farmhouse. But hopefully it can supplement and support our wine business by helping week-end and holiday visitors from all over to China to see what we are doing. Our typical customer is a young, upper-middle class couple, with or without children, who have found out about us through word-of-mouth or the internet. They usually tell me they are looking for "something different" to do with their leisure time, something "away from the crowds". Sometimes they arrive by plane, but often in smart new cars, for which this is their longest journey yet.

I think the restaurant business could prove interesting as the area develops as a "wine destination", though it will remain highly seasonal and concentrated on week-ends and public holidays. Our original chef Cai Hong has returned (hurray!). I also recently invited a Michelin-starred chef, Dr. Miguel Sanchez Romera, to cook at the castle for a short season. I cannot count the latter a financial success – I think I am about five years too early for this kind of thing – but I enjoyed watching Chinese customers' first encounter with "molecular cuisine", and the interesting opportunities for wine pairing.

Can I continue to own Treaty Port Vineyards and the castle? I hope so, though much depends upon the performance of my fund management business, as well as the needs of my family as regards health and education. The operation is no longer a financial burden – we are close to break-even – but the remote management of such a business still takes up a lot of time. Perhaps the best would be if I can find a famous wine company to buy a stake in the operation, thus relieving the pressure on management and helping with distribution.

I have learnt a lot from the project. About wine and architecture, about how the Communist Party works and about the Chinese countryside. I have got tangible satisfaction from seeing people appreciate what I have created. I have enjoyed a front row seat through an incredible period in China's development. But would I do this all again knowing what I know now?

Are you kidding?

APPENDIX I
MAP OF SHANDONG

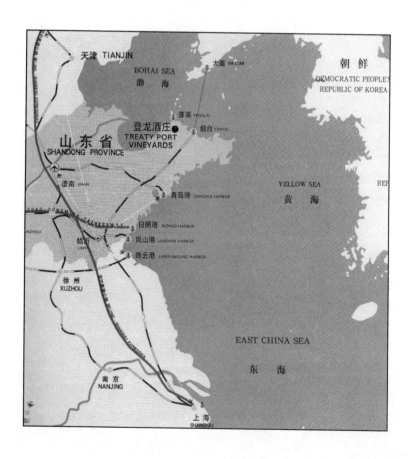

Appendix 2

Bibliography

Confucius: "Walking 10,000 miles is worth more than reading 10,000 books."

"My Thirty Years in China". Publisher Alain Charles (Chapter 6 by Chris Ruffle)

"China: an Insider's Guide" by Chris Ruffle and the Open Door team (ACA 2014)

"Scotland's Castle Culture" Edited by Audrey Deakin, Miles Glendinning and Aonghus Mackechnie. (Chapter 16 "Castles in the Modern Age" by Ian Begg) (2011)

"The Scottish Chateau" by Charles McKean

"Travels of an Alchemist". Translated by Arthur Waley (an account of Qiu Chuji's journey to see Genghis Khan written by his disciple) (1931)

"The Thistle & the Crown" by Shiona Airlie (a biography of James Stewart Lockhart: "The Commissioner") and "Scottish Mandarin" by Shiona Airlie (the Life and Times of Sir Reginald Johnston). Both spent a lot of time in Weihai.

"My Naval Career and Travels" by Admiral Sir Edward Seymour (1911). His portrait appears on our "Admiral" and "Treaty Port Port" wines.

"Bacchus in the East: the Chinese Grape Wine Industry 1892-1938" by Michael Godley (1986)

"Treaty Port Life in China 1843 – 1943" by Frances Wood "Empire Made Me" by Robert Bickers "Lilla's Feast" by Frances Osborne "China: the new Wine Frontier" by Janis Miglavs

APPENDIX 3
LABELS

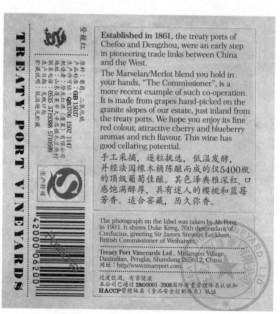

Treaty Port Vineyards Ltd.
Mujiagou Village, Daxindian, Penglai,
Shandong 265612, China
网址：http://www.treatyport.com

CASTLE 城堡白

WHITE

2009 净含量 **750** ML 酒精度 **10.5%** VOL

干白葡萄酒

CASTLE 城堡白

WHITE

2009 干白葡萄酒

Castle White is made with grapes hand-picked on the granite slopes of our estate just inland from the old treaty ports of Dengzhou and Chefoo (now Penglai and Yantai). This refreshing and spicy white is designed to complement Chinese seafood dishes.

The portrait of Elcho Castle by Renny Tait hangs at the Scottish Castle in Shandong.

城堡白是由经手工采摘的丘山花岗岩斜坡上的葡萄酿造而成，这款清香醇的白葡萄酒专为搭配中餐海鲜美食品而设计。苏各兰酒堡所在位置为旧时登州与芝采之间的通商口岸（为现今的蓬莱和烟台）。

由雷尼泰特所绘制的克雷格瓦城堡,现收藏于山东苏各兰酒堡。

本公司已经通过ISO9001-2008国际质量体系认证和HACCP管理体系（食品安全控制体系）认证。
配料清单：葡萄 食品添加剂：二氧化碳
食品标准号：GB 15037
生产许可证：Q3705 1502 1147
生产日期：见瓶封口处
产地：山东省烟台市蓬莱市
制造商：爰龙红酒（蓬莱）有限公司
地址：山东蓬莱市大辛店镇木兰沟村
联系电话：0535 5719388 5710599
贮藏说明：低温避光贮藏
过度饮酒有害健康

黑歌海娜
Grenache
Noir

桑娇维塞
Sangiovese

小白玫瑰
Muscat

TREATY PORT VINEYARDS

净含量
750ml
酒精度
12%vol

年份
VINTAGE
2014

Treaty Port Vineyards Ltd. Mulangou Village,
Daxidian, Penglai, Shandong, 265612, China
网址：http://www.treatyport.com

过量饮酒，
有害健康

THE DEBUTANTE
初阁玫瑰红

2014

2014年的丘山山谷产区，气候条件
极佳。在葡萄成熟季节没有任何的
降雨，同时光照充足，湿度低。2014年
初阁玫瑰款所采用的三种葡萄品种，
均是在这样的气候条件下成熟和采摘
的。采摘时，葡萄的糖度达到了21
度，酸度7g/L，糖酸比平衡。

葡萄汁与葡萄皮只经过了较智浸泡提
取葡萄定中的颜色，并在低温下进行
发酵，以保持葡萄的水果香气。
在发酵最止时，酒中仍然留有少量
残糖，以达到平衡酒体的目地。

这款酒具有浓郁的草莓、樱桃和玫瑰
花瓣的春气，口感圆润，甘美，清爽。
建议饮用温度10度。本款酒适合与三
文鱼，白肉和上海菜搭配，比如加三
丝，菜大黄鱼，清蒸大闸蟹和八宝鸡
等。这款酒也是起佳的餐前开胃酒。

In 2014, Qiushan Valley had perfect
weather conditions for the grape harvest.
There was no rain, sunshine hours were
long, and humidity low. This resulted in
a harvest sugar level of 21 Brix,
balanced by acidity of 7g/L.

The juice of wine was given a brief period of
skin contact to extract the necessary color.
Then the juice was fermented under low
temperature to retain its fruity flavors.
The ferment was stopped when a small
quantity of residual sugar was left to
balance the finished wine.

This wine is full of strawberry, cherry
and rose petal aromas; the palate is well
rounded, fresh and luscious. Serve chilled
and enjoy with salmon, white meat and
Shanghai dishes. Also good as an aperitif.

The image on the front label
shows young Chinese socialites
celebrating in the 1930s.

配料清单：葡萄 食品添加剂：二氧化硫
产品类型：半干型
产品标准：GB 15037
生产许可证：QS3706 1502 1147
制造者：登龙红酒（蓬莱）有限公司
地址：山东蓬莱市大辛店镇木兰沟村
联系电话：0535 5719388 5710599
生产日期：见瓶口处
产地：山东省烟台市
规格：750ml/瓶
保质期：10年
贮藏说明：低温避光贮藏

6 951765 300161

In 2014, Qiushan Valley had perfect weather conditions for the grape harvest. There was no rain, sunshine hours were long, and humidity low. This resulted in a harvest sugar level of 23 Brix, balanced by acidity of 6.5g/L. This is a limited release of only 5000 bottles.

This wine displays all the hallmarks of an exceptional vintage in Penglai. It presents a swathe of dark cherry, blueberry, cassis, cedar and dry tobacco leaf in the bouquet. On the palate, this wine shows an alluring complexity and delicacy with a long finish. Enjoy with cheese, dark chocolate, braised pork and Xinjiang roast lamb.

This photograph of Prince Gong was taken by Scottish photographer John Thomson in 1872. Prince Gong acted as the Prince Regent from 1861 to 1884, and was a proponent of friendly relations with the West and China's modernization. He established the Zhongli Yamen, the progenitor of China's Foreign Ministry, in 1861, and the Tongwen Guan, which became Beijing University, in 1862.

6 951765 300178